SpringerBriefs in Mathematics

For further volumes:
http://www.springer.com/series/10030

Vladimir Rovenski • Paweł Walczak

Topics in Extrinsic Geometry of Codimension-One Foliations

 Springer

Vladimir Rovenski
Department of Mathematics
University of Haifa
31905 Haifa, Israel
rovenski@math.haifa.ac.il

Paweł Walczak
Department of Mathematics
University of Lodz
90238 Lodz, Poland
pawelwal@math.uni.lodz.pl

ISSN 2191-8198   e-ISSN 2191-8201
ISBN 978-1-4419-9907-8   e-ISBN 978-1-4419-9908-5
DOI 10.1007/978-1-4419-9908-5
Springer New York Dordrecht Heidelberg London

Library of Congress Control Number: 2011931089

Mathematics Subject Classification (2010): 53C12, 35L45, 26B20

Printed on acid-free paper

Springer is part of Springer Science+Business Media (www.springer.com)

V. Rovenski dedicates this book to his parents:
Ira Rushanova and Yuzef Rovenski

P. Walczak dedicates this book to his family.

Foreword

The authors of this work asked me to read it and write a foreword. I did so with pleasure because differential geometry of foliations was one of my research subjects decades ago.

Foliations, i.e., partitions into submanifolds of a constant, lower dimension, are beautiful structures on manifolds that encode a lot of geometric information. The topological study of foliations was initiated by Ch. Ehresmann and G. Reeb in the 1940s and soon became a research subject of many mathematicians. In particular, the study of the smooth case and of the differential geometric aspects became an important part of foliation theory, developed in the early stages by B. Reinhart, R. Bott, F. Kamber, Ph. Tondeur, P. Molino, and many others.

The present work is a research monograph and is addressed to readers who have enough knowledge of differential and Riemannian geometry. Its first two chapters are devoted to the development of a computational machinery that provides integral and variational formulas for the most general, extrinsic invariants of the leaves of a foliation of a Riemannian manifold. The third chapter defines a very general notion of extrinsic geometric flow and studies the evolution of the leaf-wise Riemannian metric along the trajectories of this flow. The authors give existence theorems and estimations of the maximal evolution time and make a study of soliton solutions.

The authors of the present monograph are well known specialists in the field, with previously published books and papers on the differential geometry of foliations of Riemannian manifolds. Here, they succeed a technical tour de force, which will lead to important geometric results in the future and I recommend this work to all those who have an interest in the differential geometry of submanifolds and foliations of Riemannian manifolds. They will find methods and results that bring profit to their research.

University of Haifa, Israel   Izu Vaisman

Preface

The subject and the history. Foliation theory is about 60 years old. The notion of a foliation appeared in the 1940s in a series of papers of G. Reeb and Ch. Ehresmann, culminating in the book [40]. Since then, the subject has enjoyed a rapid development. Foliations relate with such topics as vector fields, integrable distributions, almost-product structures, submersions, fiber bundles, pseudogroups, Lie groups actions, and explicit constructions (Hopf and Reeb foliations).

Reeb also published a paper [41] on extrinsic geometry of foliation in which he proved that the integral of the mean curvature of the leaves of any codimension-one foliation on any closed Riemannian manifold equals zero. By *extrinsic geometry* we mean properties of foliations on Riemannian manifolds which can be expressed in terms of the second fundamental form of the leaves and its invariants (principal curvatures, scalar mean curvature, higher mean curvatures, and so on).

More precisely, if $\mathscr{F}$ is a smooth foliation of a Riemannian manifold (M, g) then the *second fundamental forms* B_L of all the leaves $\{L\}$ of $\mathscr{F}$ provide a vector-valued symmetric tensor B on M defined by:

$$B(X, Y) = (\nabla_X Y)^{\perp},$$

where ∇ is the Levi-Civita connection on (M, g), X and Y are tangent to $\mathscr{F}$, and $(\cdot)^{\perp}$ denotes the projection of the tangent bundle TM onto the orthogonal complement $T^{\perp}\mathscr{F}$ of the bundle $T\mathscr{F}$ consisting of all the vectors tangent to (the leaves of) $\mathscr{F}$. The tensor B can be extended to the whole tangent bundle of M by $B(N, \cdot) = 0$ whenever N is orthogonal to $\mathscr{F}$. If $\mathscr{F}$ is of codimension 1 and transversely oriented, B induces a symmetric scalar $(0,2)$-tensor field b (the *second fundamental form*) given by

$$b(X, Y) = g(B(X, Y), N)$$

for all X and Y. All the properties of $\mathscr{F}$ which can be expressed in terms of B (respectively, b) belong to extrinsic geometry. For example, a foliation $\mathscr{F}$ is called *totally geodesic* when $B \equiv 0$,

minimal when the mean curvature vector $H = \frac{1}{n} \operatorname{Tr}_g(B)$ of $\mathscr{F}$ vanishes,

umbilical when $B(X,Y) = H \cdot g(X,Y)$ for all $X, Y \in T\mathscr{F}$, and so on.

One of the principal problems of extrinsic geometry of foliations reads as follows: *Given a foliation $\mathscr{F}$ on a manifold M and an extrinsic geometric property (P), does there exist a Riemannian metric g on M such that $\mathscr{F}$ enjoys (P) with respect to g?*

Similarly, one may ask the following, analogous question:

Given a manifold M and an extrinsic geometric property (P), does there exist a foliation $\mathscr{F}$ and a Riemannian metric g on M such that $\mathscr{F}$ enjoys (P) with respect to g?

Such problems (first posed by H. Gluck for geodesic foliations) were studied already in the 1970s when Sullivan [50] provided a topological condition (called *topological tautness*) for a foliation, equivalent to *geometrical tautness*, that is existence of a Riemannian metric making all the leaves minimal. From classical theorem of Novikov [32] and results of Sullivan, it follows directly that the three-dimensional sphere S^3 admits no two-dimensional foliations which are minimal with respect to any Riemannian metric. For instance, there is no metric making a Reeb foliation $\mathscr{F}_R$ on a three-dimensional sphere minimal.

Umbilizable foliations on M^3 are transversely holomorphic, hence, see [11]: *If a closed orientable M^3 admits an umbilical foliation then it is diffeomorphic to the total space of a Seifert fibration (all one-dimensional leaves are closed) or of a torus bundle over the circle.* For example, since S^3 is the total space of a Seifert fibration, there exist metrics making a Reeb foliation $(S^3, \mathscr{F}_R)$ umbilical. Another example of this type may be found in a recent paper by Langevin and the second author [30]: closed Riemannian spaces of negative Ricci curvature admit no codimension-1 umbilical foliations.

In recent decades, several tools providing results of this sort have been developed. Among them, one may find Sullivan's [49] *foliated cycles* and several *Integral Formulae* ([3, 9, 45, 46, 54], etc.), the very first of which is G. Reeb's vanishing of the integral of the mean curvature mentioned earlier.

The authors also have been interested in extrinsic geometry of foliations for a long time (see, for example, [42–44, 54–58]) and this work is, in some sense, a continuation of this interest.

The contents. The book includes several topics in Extrinsic Geometry of Foliations. The first topic presented in the book (Chap. 1) is a series of new Integral Formulae, for a codimension-one foliation on a closed Riemannian manifold. The formulae depend on the Weingarten operator, the Riemannian curvature tensor (e.g., Jacobi operator), and their scalar invariants. Integral formulae begin with the classic formula by Reeb, for manifolds of constant curvature they reduce, to known formulae by Brito et al. [9], and Asimov [4]. Integral formulae can be useful for the following problems: prescribing higher mean curvatures (or other symmetric functions of principal curvatures) of foliations; minimizing volume and energy defined for vector or plane fields on manifolds; existence of foliations whose leaves enjoy a given geometric property such as being totally geodesic, umbilical, minimal, etc.

The central topic of the book is *Extrinsic Geometric Flow* (*EGF*, for short, see Chap. 3) on foliated manifolds $(M, \mathscr{F})$, $\mathrm{codim}\, \mathscr{F} = 1$, which may provide more results on geometry of foliations. EGFs arise as solutions to the partial differential equation (PDE)

$$\partial_t g_t = h(b_t),$$

where (g_t), $t \in [0, T)$, are Riemannian metrics on M along the leaves and $h(b_t)$ the symmetric $(0, 2)$-tensors along the leaves expressed in terms of the second fundamental form b_t of $\mathscr{F}$ on (M, g_t); $h(b_t)$ being identically zero in the direction orthogonal to $\mathscr{F}$. In particular, EGF – for suitable choice of the right-hand side in the EGF equation – may provide families (g_t) of Riemannian structures on a given foliated manifold $(M, \mathscr{F})$ converging as $t \to T$ to a metric g_T for which $\mathscr{F}$ satisfies a given geometric property (P), say, is umbilical, minimal, or just totally geodesic.

A *Geometric Flow* is an evolution of a given geometric structure under a differential equation associated to a functional on a manifold which has geometric interpretation, usually associated with some (either extrinsic or intrinsic) curvature. Geometric flows play an essential role in many fields of mathematics and physics. They all correspond to dynamical systems in the infinite dimensional space of all possible geometric structures (of given type) on a given manifold.

The strong interest of scientists in GF of various types is demonstrated by Annual International Workshops (*GF in Mathematics and Physics*, 2006 – 2011, BIRS Banff; *GF in finite or infinite dimension*, 2011, CIRM; *Geometric Evolution Equations*, 2011, University of Constance; *GF and Geometric Operators*, 2009, Centro De Giorgi, Pisa, and so on).

To some extent, the idea of EGF is analogous to that of the famous *Ricci flow*. In the Ricci flow equation, the configuration space is a single manifold and the Riemannian structures are deformed by quantities which belong to intrinsic geometry, in the case of EGFs, the configuration space is a foliated manifold while the Riemannian structures are deformed by invariants of extrinsic geometry. In both cases, the (EGF or Ricci flow) equation makes sense because both its sides are symmetric tensors of the same type. Notice that the study of the Ricci flow provided the proof of outstanding conjectures: Poincaré Conjecture and Thurston Geometrization Conjecture.

To apply EGF to various problems of extrinsic geometry, one needs *variational formulae* (see Chap. 2) which express variation of different quantities belonging to extrinsic geometry of a fixed foliation under variation of the Riemannian structure of the ambient manifold. Also, some special solutions (called *extrinsic geometric solitons* here, EGS, for short, see Sect. 3.8) of the EGF equation are of great interest because, in several cases, they provide Riemannian structures with very particular geometric properties of the leaves.

Throughout the book, (M^{n+1}, g_t) is a Riemannian manifold with a codimension one transversely oriented foliation $\mathscr{F}$, ∇^t the Levi-Civita connection of g_t,

$$2g_t(\nabla^t_X Y, Z) = X(g_t(Y,Z)) + Y(g_t(X,Z)) - Z(g_t(X,Y))$$
$$+ g_t([X,Y],Z) - g_t([X,Z],Y) - g_t([Y,Z],X)$$

for all the vector fields X, Y, Z on M, N the positively oriented unit normal to $\mathscr{F}$ with respect to any g_t, $A : X \in T\mathscr{F} \mapsto -\nabla^t_X N$ the Weingarten operator of the leaves, which we extend to a $(1,1)$-tensor field on TM by $A(N) = 0$.

Observe that the difference of two connections is always a tensor, hence $\Pi_t := \partial_t \nabla^t$ is a $(1,2)$-tensor field on (M, g_t). Differentiating with respect to t the above classical formula yields the known formula, which allows us to express Π_t by:

$$2g_t(\Pi_t(X,Y),Z) = (\nabla^t_X S)(Y,Z) + (\nabla^t_Y S)(X,Z) - (\nabla^t_Z S)(X,Y),$$

where $S = \partial_t g_t$ is time-dependent symmetric $(0,2)$-tensor field and $X, Y, Z \in TM$.

The definition of the $\mathscr{F}$-truncated (r,k)-tensor field $\hat{S}$ (where $r = 0, 1$, and $\hat{}$ denotes the $T\mathscr{F}$-component) will be helpful in Chaps. 2 and 3,

$$\hat{S}(X_1, \ldots, X_k) = S(\hat{X}_1, \ldots, \hat{X}_k) \qquad (X_i \in TM).$$

Acknowledgments The authors would like to thank their colleagues, David Blanc and Izu Vaisman (Mathematical Department, University of Haifa), Krzysztof Andrzejewski, Wojciech Kozłowski, Kamil Niedziałomski and Szymon Walczak (Faculty of Mathematics and Computer Science, University of Łódź) for helpful corrections concerning the manuscript. The authors warmly thank Elizabeth Loew and Ann Kostant for support in the publishing process. The authors are also greatly indebted to Marie-Curie actions support of their research by grants EU-FP7-PEOPLE-2007-IEF, No. 219696 and EU-FP7-PEOPLE-2010-RG, No. 276919.

Haifa – Łodz   Vladimir Rovenski
                                                                     Paweł Walczak

Contents

Acronyms

EGF	Extrinsic Geometric Flow
EGS	Extrinsic Geometric Soliton
VF	Variation Formula
IF	Integral Formula
g_t	t-dependent Riemannian metric
(M, g_t)	Riemannian manifold (with a t-dependent metric)
$\mathscr{F}, L_\alpha$	A codimension-one foliation and its leaves
N	Unit normal to (the leaves of) a foliation $\mathscr{F}$
A, b	Weingarten operator and 2-nd fundamental form for $\mathscr{F}$ with respect to N
$\hat{g}$	$\mathscr{F}$-truncated metric tensor
$g^\perp$	N-component of the metric g
$\hat{b}_j$	$\mathscr{F}$-truncated tensor dual to A^j
$h(b)$	The symmetric $(0,2)$ tensor expressed in terms of b
∇^t	The Levi-Civita connection for g_t
σ_k	k-th elementary symmetric function of A
τ_k	k-th power sum (symmetric function) of the eigenvalues of A
$\vec{\tau}$	The vector function $(\tau_1, \ldots, \tau_n)$
Γ_{ij}^k	Christoffel symbols for the metric g
$\flat$	the (musical) isomorphism $\flat : TM \to T^*M$, i.e., $X^\flat = g(X, \cdot)$
$\sharp$	The (musical) isomorphism $\sharp : T^*M \to TM$
λ	The normal curvature of an umbilical foliation
k_i	The principal curvatures of the leaves
$\Lambda_l^k(M)$	The bundle of $\mathscr{F}$-truncated (k,l)-tensors on (M, g)
$\langle \, , \, \rangle$	The inner product of tensors
$\mathscr{M}(M, \mathscr{F}, N)$	The space of Riemannian metrics on M of finite volume with N being a unit normal to $\mathscr{F}$
R, Ric	The Riemannian curvature and the Ricci tensors
R_N	$= R(\cdot, N)N$ – the Jacobi operator. Indeed, $\operatorname{Tr} R_N = \operatorname{Ric}(N, N)$

Chapter 1
Integral Formulae

Abstract The chapter presents a series of new *Integral Formulae* (IF) for a codimension-one foliation on a closed Riemannian manifold. The proof of IF is based on the Divergence Theorem. The IF start from the formula by Reeb, for foliations on space forms they generalize the classical ones by Asimov, Brito, Langevin, and Rosenberg. Our IF include also a set of arbitrary functions f_j depending on the scalar invariants of the Weingarten operator. For a special choice of auxiliary functions the IF involve the Newton transformations of the Weingarten operator. We apply IF to umbilical foliations and foliations whose leaves have constant second-order mean curvature.

1.1 Introduction

Here, for the readers' convenience, we provide the following standard definition from foliation theory, see [12].

Definition 1.1. A family $\mathscr{F} = \{L_\alpha\}_{\alpha \in A}$ of connected subsets of a manifold M^m is said to be an *n-dimensional foliation*, Fig. 1.1, if:

(1) $\bigcup_{\alpha \in A} L_\alpha = M^m$.
(2) $\alpha \neq \beta \Rightarrow L_\alpha \cap L_\beta = \emptyset$.
(3) For any point $q \in M$ there exists a C^r-chart (local coordinate system) $\varphi_q : U_q \to \mathbb{R}^m$ such that $q \in U_q$, $\varphi_q(q) = 0$, and if $U_q \cap L_\alpha \neq \emptyset$ the connected components of the sets $\varphi_q(U_q \cap L_\alpha)$ are given by equations $x_{n+1} = c_{n+1}, \ldots, x_m = c_m$, where c_j's are constants. The sets L_α are immersed submanifolds of M called *leaves* of $\mathscr{F}$.

The family of all the vectors tangent to the leaves is the integrable subbundle of TM denoted by $T\mathscr{F}$. If M carries a Riemannian structure, $T\mathscr{F}^\perp$ denotes the subbundle of all the vectors orthogonal to the leaves. A foliation $\mathscr{F}$ is said to be *orientable* (respectively, *transversely orientable*) if the bundle $T\mathscr{F}$ (respectively, $T\mathscr{F}^\perp$) is orientable.

V. Rovenski and P. Walczak, *Topics in Extrinsic Geometry of Codimension-One Foliations*, SpringerBriefs in Mathematics, DOI 10.1007/978-1-4419-9908-5_1,
© Vladimir Rovenski and Paweł Walczak 2011

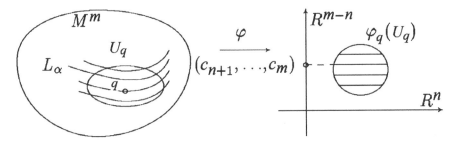

Fig. 1.1 Foliation

From the definition earlier, it follows that the bundle $T\mathscr{F}$ is *involutive*, that is the *Lie bracket* $[X,Y]$ of two sections of $T\mathscr{F}$ is a section of $T\mathscr{F}$ again. Let us recall that the classical *Frobenius Theorem* says that any *distribution* (that is, a subbundle of TM) D on M is involutive if and only if $D = T\mathscr{F}$ for some foliation $\mathscr{F}$. Clearly, any one-dimensional distribution is involutive, so tangent to a one-dimensional foliation.

The [12, I, Part 2] is a short course on the codimension-one foliations. The [12, II, Part 3] studies compact 3-manifolds foliated by surfaces, a popular topic since the theorem by Novikov on the existence of Reeb components for foliated 3-sphere. Taut foliations are used as powerful tools for studying 3-manifold topology. Due to Thurston and Gabai, a *taut foliation* is a codimension-one foliation with the property that there is a single transverse circle intersecting every leaf. By a result of D. Sullivan, see [50], a codimension-one foliation is taut if there exists a Riemannian metric that makes each leaf a minimal surface.

Let (M,g) be a Riemannian manifold with the metric g and the Levi-Civita connection ∇. In particular, we have $\nabla g = 0$. The manifold M will be always *closed* (i.e., compact and without boundary). The *Riemannian curvature tensor* is given by:

$$R(X,Y)V = (\nabla_X\nabla_Y - \nabla_Y\nabla_X - \nabla_{[X,Y]})V, \quad X,Y,V \in \Gamma(TM).$$

Its trace (with respect to the metric g) is called the *Ricci curvature tensor*. If (e_i) are local orthonormal vector fields on M then

$$\mathrm{Ric}(X,Y) = \sum_i g(R(e_i,X)Y, e_i), \quad X,Y \in \Gamma(TM).$$

The trace, $\mathrm{Scal} = \sum_i \mathrm{Ric}(e_i, e_i)$, is called the *scalar curvature*. One of the tools to describe the Riemannian curvature is the *sectional curvature*, $K_\sigma = g(R(X,Y)Y,X)$, where X,Y is the orthonormal basis of a two-dimensional subspace $\sigma \subset TM$. It is the Gaussian curvature of surfaces. If M is a space of constant curvature c then

$$R(X,Y)V = c(g(Y,V)X - g(X,V)Y), \qquad K_\sigma = c.$$

Let N be a unit normal vector field of a transversally oriented codimension-one foliation $\mathscr{F}$. Indeed, $\mathrm{Ric}(N,N) = \mathrm{Tr}\,R_N$, where $R_N = R(\cdot,N)N$ is the Jacobi operator.

By *integral formula* (IF) we mean the vanishing of the integral over M of an expression depending on the Weingarten operator of $\mathscr{F}$, the Riemannian curvature tensor of (M, g), and their scalar invariants.

The *total* higher-order mean curvatures and power sums are the integrals

$$I_{\sigma,j} = \int_M \sigma_j \, d\,\mathrm{vol}, \quad I_{\tau,j} = \int_M \tau_j \, d\,\mathrm{vol}, \quad j = 1, 2, \ldots \qquad (1.1)$$

see the definition of σ_j and τ_j in Section 1.2.2.

Coming back to the main topic of this chapter, let us recall that the first known integral formula (for codimension-one foliations) belongs to Reeb [41],

$$\int_M H \, d\,\mathrm{vol} = 0, \qquad (1.2)$$

where H is the mean curvature of leaves. By (1.2) we have $I_{\tau,1} = I_{\sigma,1} = 0$. The proof of (1.2) is based on the Divergence Theorem, and the identity $\mathrm{div} N = -nH$. Consequently, if a foliation has a projectable mean curvature function then, according to (1.2), one of its leaves has mean curvature equal to zero, so this leaf is minimal. Also, (1.2) and its counterparts for foliated domains with boundary provide the only obstructions for a function f on a compact foliated manifold $(M, \mathscr{F})$ to become the mean curvature σ_1 of $\mathscr{F}$ with respect to some Riemannian metric on M. The conditions which are necessary and sufficient in this case read [33]: either $f \equiv 0$ or there exist points $x, y \in M$ for which $f(x)f(y) < 0$, and, in any positive (respectively, negative) foliated domain $D \subset M$ there exists a point x such that $f(x) > 0$ (respectively, $f(x) < 0$).

The next well-known integral formula in the series for total σ's is

$$\int_M \left(2\,\sigma_2 - \mathrm{Ric}(N, N) \right) d\,\mathrm{vol} = 0. \qquad (1.3)$$

For $n = 1$ and $\dim M = 2$, we have $\sigma_2 = 0$, and (1.3) reduces to the integral of Gaussian curvature, $\int_M K \, d\,\mathrm{vol} = 0$, and means just that the Euler characteristic of M^2 with a unit vector field N is zero. The formula (1.3) follows from a result in [54] (for the foliation case, see also [39]) and has many applications. For example, one can see directly that it implies nonexistence of umbilical foliations on closed manifolds of negative curvature: if the integrand is strictly positive, so is the value of the integral. In [30], (1.3), together with some standard tools of analysis (Hölder inequality and so on), was used to show that codimension-one foliations of closed negatively Ricci-curved manifolds are "far" (in some sense) from being umbilical.

Brito et al. [9], extending the result by D. Asimov [4] for gaussian curvature, have shown that the integrals $I_{\sigma,j}$ on a compact space form $M^{n+1}(c)$ do not depend on $\mathscr{F}$: they depend on n, j, c and the $\mathrm{vol}(M)$ only,

$$I_{\sigma,j} = \begin{cases} c^{j/2} \binom{n/2}{j/2} \mathrm{vol}(M), & n \text{ and } j \text{ even} \\ 0, & \text{either } n \text{ or } j \text{ odd.} \end{cases} \qquad (1.4)$$

Example 1.1. Here is an amazing consequence of (1.4) for any sufficiently smooth codimension-one foliation on the round unit sphere S^3 (communicated to the authors by D. Asimov in 2008). By S. Novikov's theorem, any such foliation contains a leaf diffeomorphic to a torus. So, by Gauss-Bonnet theorem, there is a point with zero *Gaussian leaf curvature* $K_{\mathscr{F}}$. By (1.4) for $n = j = 2$ and $c = 1$, the average of σ_2 is 1. Because $K_{\mathscr{F}} = 1 + \sigma_2$, there is a point, where $K_{\mathscr{F}} > 1 + 1 = 2$. Hence, the set of values of the function $K_{\mathscr{F}} : S^3 \to \mathbb{R}$ contains the interval $[0, 2 + \varepsilon]$ for some $\varepsilon > 0$.

In this direction, following [3] and [44], we present a series of new IF (for symmetric functions σ's and τ's), which start from (1.2) and (1.3). For manifolds of constant curvature the IF are reduced (by a simple choice of functions f_j) to (1.4).

IF are useful for several problems: prescribing mean curvatures σ_j (or other symmetric functions of the principal curvatures k_i) of foliations, minimizing volume and energy defined for vector fields on manifolds, and existence of foliations whose leaves enjoy a given geometric property such as being minimal, umbilical, etc. (see, e.g., [3], [16], [42], [46], [47], [52], [54] and the bibliographies therein).

1.2 Preliminaries

We shall use the *Divergence Theorem* $\int_M \operatorname{div}(f) \, d\operatorname{vol} = 0$ and the identity

$$\operatorname{div}(fX) = f \operatorname{div} X + X(f)$$

for smooth functions f and vector fields X on M. In some cases we calculate the leaf-wise divergence $\operatorname{div}_{\mathscr{F}}$ (along $\mathscr{F}$). The trace of $(1,1)$-tensors will be calculated along $\mathscr{F}$. We denote by $^\perp$ and $^\top$ the normal and tangent to $\mathscr{F}$ components of vectors, respectively. In what follows, we briefly write

$$\nabla_N N = Z.$$

1.2.1 Hypersurfaces of Riemannian Manifolds

When we have an isometric immersion (of manifolds) from L into M, we say that L is an immersed submanifold of M. If, in addition, the inclusion is an embedding then L is said to be an *embedded submanifold* of M. At each point $q \in L$, the inner product $g(\cdot, \cdot)$ on $T_q M$ induces an inner product on $T_q L$ that we call the induced Riemannian metric and denote it by the same symbol. We will consider always submanifold of a Riemannian manifold with the metric that is induced in this way.

We decompose $\nabla_X Y$, where $X, Y \in \Gamma(TL)$, into its tangent part $(\nabla_X Y)^\top$ and its normal part $(\nabla_X Y)^\perp$. Then the *Levi-Civita connection* ∇^L of L is given by $\nabla^L_X Y = (\nabla_X Y)^\top$, and one defines the second fundamental tensor of L by $B(X, Y) = (\nabla_X Y)^\perp$.

We shall consider codimension-one (transversally oriented) submanifolds, called *hypersurfaces*. Let N be a unit positive oriented normal vector field of L. We call b defined by:

$$b(X,Y) = g(B(X,Y),N)$$

the scalar *second fundamental form*. This gives the orthogonal decomposition:

$$\nabla_X Y = \nabla_X^L Y + B(X,Y) = \nabla_X^L Y + b(X,Y)N,$$

called the *Gauss formula*. The tensor field $A = -(\nabla_X N)^\top$ is called the *Weingarten (shape) operator* of L (in direction N), and is related to b by $b(X,Y) = g(AX,Y)$. The symmetry of B implies that A is self-adjoint. The orthogonal decomposition

$$\nabla_X^L N = (\nabla_X N)^\perp - AX$$

is known as the *Weingarten formula*. The formulae of Gauss and Weingarten can be seen as first-order differential equations. The covariant derivatives of the second fundamental form and of the Weingarten operator are given by the formulas:

$$(\nabla_X^\perp B)(Y,V) = \nabla_X^\perp B(Y,V) - B(\nabla_X Y,V) - B(Y,\nabla_X V),$$
$$(\nabla_X A)Y = \nabla_X(AY) - A(\nabla_X Y).$$

Using the formulae of Gauss and Weingarten, one can obtain that

$$(R(X,Y)V)^\perp = (\nabla_X^\perp B)(Y,V) - (\nabla_Y^\perp B)(X,V),$$
$$(R(X,Y)V)^\top = R^L(X,Y)V + g(AX,V)AY - g(AY,V)AX. \tag{1.5}$$

The first equation of (1.5) is called the *Gauss equation*, the second one the *Codazzi equation*. Notice that $(R(X,Y)V)^\perp = 0$ when M is a space of constant curvature.

The difference $\operatorname{Rm}^{\mathrm{ex}}(X,Y)V = R^L(X,Y)V - (R(X,Y)V)^\top$ is called the *extrinsic Riemannian curvature* of L, see Sect. 3.9.1. The *extrinsic sectional, Ricci,* and *scalar curvatures* are given, respectively, by:

$$K^{\mathrm{ex}}(X,Y) = K^L(X,Y) - K(X,Y) = g(AX,X)g(AY,Y) - g(AX,Y)^2,$$
$$\operatorname{Ric}^{\mathrm{ex}}(X,Y) = g(AX,AY) - (\operatorname{Tr} A)g(AX,Y),$$
$$\operatorname{Scal}^{\mathrm{ex}} = \operatorname{Tr} \operatorname{Ric}^{\mathrm{ex}} = (\operatorname{Tr} A)^2 - \operatorname{Tr}(A^2).$$

1.2.2 Invariants of the Weingarten Operator

Power sums of the principal curvatures $k_1,\dots,k_n$ (the eigenvalues of A) are given by:

$$\tau_j = k_1^j + \dots + k_n^j = \operatorname{Tr}(A^j), \quad j \geq 0.$$

The τ's can be expressed using the *elementary symmetric functions* $\sigma_1, \ldots, \sigma_n$

$$\sigma_j = \sum_{i_1 < \ldots < i_j} k_{i_1} \cdot \ldots \cdot k_{i_j} \quad (0 \leq j \leq n),$$

called *mean curvatures* in the literature. Notice that $\sigma_0 = 1$, $\sigma_n = \det A$, and $\tau_0 = n$.

Remark 1.1. Evidently, the functions τ_{n+i} $(i > 0)$, are not independent: they can be expressed as polynomials of $\vec{\tau} = (\tau_1, \ldots, \tau_n)$, using the *Newton formulae*

$$\tau_j - \tau_{j-1}\sigma_1 + \ldots + (-1)^{j-1}\tau_1\sigma_{j-1} + (-1)^j j\sigma_j = 0 \quad (1 \leq j \leq n),$$
$$\tau_j - \tau_{j-1}\sigma_1 + \ldots + (-1)^n \tau_{j-n}\sigma_n = 0 \quad (j > n),$$

which in matrix form are:

$$T_n \begin{pmatrix} \sigma_1 \\ \sigma_2 \\ \ldots \\ \sigma_n \end{pmatrix} = \begin{pmatrix} \tau_1 \\ \tau_2 \\ \ldots \\ \tau_n \end{pmatrix}, \text{ where } T_n = \begin{pmatrix} 1 & & & & \\ \tau_1 & -2 & & & \\ \ldots & \ldots & \ldots & \ldots & \\ \tau_{n-1} & -\tau_{n-2} & \ldots & (-1)^n \tau_1 & (-1)^{n+1} n \end{pmatrix}.$$

Hence, the σ's can be expressed in terms of the τ's using T_n^{-1}. Moreover, we have

$$\tau_i = \det \begin{pmatrix} \sigma_1 & 1 & 0 & \ldots & \ldots \\ 2\sigma_2 & \sigma_1 & 1 & 0 & \ldots \\ \ldots & \ldots & \ldots & \ldots & \\ i\sigma_i & \sigma_{i-1} & \ldots & \sigma_2 & \sigma_1 \end{pmatrix}, \quad i = 1, 2, \ldots$$

These formulae can be used to express the τ_{n+i}'s as polynomials of $\tau_1, \ldots, \tau_n$.

Many authors investigated recently higher-order mean curvatures of hypersurfaces using the Newton transformations of the shape operator. Newton transformations of the shape operator have been applied successfully to foliations, see [3, 44] and [47]. *Newton transformations* $T_r(A)$ are defined either inductively by:

$$T_0(A) = \widehat{\text{id}}, \quad T_r(A) = \sigma_r \widehat{\text{id}} - A \cdot T_{r-1}(A), \quad 1 \leq r \leq n,$$

or explicitly as:

$$T_r(A) = \sum_{i=0}^r (-1)^i \sigma_{r-i} A^i = \sigma_r \widehat{\text{id}} - \sigma_{r-1} A + \ldots + (-1)^r A^r, \quad 0 \leq r \leq n. \quad (1.6)$$

Notice that A and $T_r(A)$ commute. By the Cayley-Hamilton Theorem, $T_n(A) = 0$.

Let $\lambda = (\lambda_1, \ldots, \lambda_m) \in \mathbb{Z}_+^m$. The *generalized mean curvatures* $\sigma_\lambda(A_1, \ldots A_m)$ of a set of $n \times n$ matrices $A_1, \ldots A_m$ are defined in [45] by:

$$\det(I_n + t_1 A_1 + \ldots + t_m A_m) = \sum_{\lambda_1 + \ldots + \lambda_m \leq n} \sigma_\lambda(A_1, \ldots A_m) t_1^{\lambda_1} \cdot \ldots \cdot t_m^{\lambda_m}. \quad (1.7)$$

For example, when $m = 1$, we have $\det(I_n + tA) = \sum_{r \leq n} \sigma_r(A) t^r$.

Lemma 1.1 (see [47]). *If $B(t)$ $(t \geq 0)$ is a smooth family of $n \times n$ matrices then*

$$\dot{\sigma}_r(B) = \sigma_{1,r-1}(\dot{B}, B) \quad \text{(for all } r \leq n\text{).} \tag{1.8}$$

Proof. Let $B(t) = B + t\dot{B} + o(t)$, hence

$$\sum_{r \leq n} \sigma_r(B(t)) s^r = \det(I_n + sB(t)) = \det(I_n + s(B + t\dot{B})) + o(t).$$

By the definition (1.7), we have

$$\det(I_n + s(B + t\dot{B})) = \det(I_n + st\dot{B} + sB)$$
$$= \sum_{r \leq n} \sigma_r(B) s^r + \sum_{r < n} \sigma_{1,r-1}(\dot{B}, B) s^r t + o(t).$$

From the above (1.8) follows. $\qquad\square$

Lemma 1.2 (see [45]). *For arbitrary $n \times n$ matrices B, C and $r, l > 0$ we have*

$$\sigma_{r,l}(B, C) = \sigma_r(B)\,\sigma_l(C) - \sum_{i=1}^{\min(r,l)} \sigma_{r-i,l-i,i}(B, C, BC). \tag{1.9}$$

In particular,

$$\sigma_{r,1}(B, C) = \sum_{i=0}^{r}(-1)^i \sigma_{r-i}(B)\,\mathrm{Tr}\,(B^iC) = \mathrm{Tr}\,(T_{r-1}(B)\,C). \tag{1.10}$$

Proof. Comparing the determinant $\det(I_n + tB + sC + ts\,BC)$ with the product of the determinants $\det(I_n + tB)\det(I_n + sC)$ we get the equality

$$\sum_{r,l=1}^{n} \sigma_r(B)\,\sigma_l(C)\,t^r s^l = \sum_{a,b,c=1}^{n} \sigma_{a,b,c}(B, C, BC)\,t^a s^b (st)^c$$
$$= \sum_{r,l=1}^{n} t^r s^l \left(\sigma_{r,l}(B, C) + \sum_{i=1}^{\min(r,l)} \sigma_{r-i,l-i,i}(B, C, BC)\right),$$

which implies relation (1.9). From (1.9) with $l = 1$ we conclude that $\sigma_{r,1}(B, C) = \sigma_r(B)\,\sigma_1(C) - \sigma_{r-1,1}(B, BC)$. From the above, by induction, we obtain (1.10). $\quad\square$

Remark 1.2. Let $B(t)$ be a smooth family of n-by-n matrices with the symmetric functions $\tau_j = \mathrm{Tr}\,B^j$. Using the identity $\dot{B}^j = B\dot{B}^{j-1} + \dot{B}B^{j-1}$ for $j > 1$, by induction we find $\dot{B}^j = \sum_{i=1}^{j} B^{i-1}\dot{B}B^{j-i}$. By the property $\mathrm{Tr}\,(AB) = \mathrm{Tr}\,(BA)$ and the fact that for matrices the trace commutes with derivatives, we conclude that

$$\dot{\tau}_j(B) = \mathrm{Tr}\,(\dot{B}^j) = j\,\mathrm{Tr}\,(B^{j-1}\dot{B}). \tag{1.11}$$

By (1.8) and (1.10), the elementary symmetric functions of matrices $B(t)$ satisfy

$$\dot{\sigma}_r = \mathrm{Tr}\,(T_{r-1}(B)\dot{B}), \quad r = 1, 2, \ldots, n. \tag{1.12}$$

Lemma 1.3 (see [3]). *For the Weingarten operator A we have*

$$\operatorname{Tr} T_r(A) = (n - r)\,\sigma_r,$$

$$\operatorname{Tr}(A \cdot T_r(A)) = (r + 1)\,\sigma_{r+1},$$

$$\operatorname{Tr}(A^2 \cdot T_r(A)) = \sigma_1\,\sigma_{r+1} - (r + 2)\,\sigma_{r+2},$$

$$\operatorname{Tr}(T_{r-1}(A)(\nabla_X^{\mathscr{F}} A)) = X(\sigma_r), \quad X \in \Gamma(D).$$

Proof. The first three algebraic properties follow from Newton formulae of Remark 1.1. The last identity follows from (1.12). □

In this chapter, we shall consider also integrals of the quantities involving the vector field $h(A)Z$, where h is of the form described later. Let b_j be a symmetric $(0,2)$-tensor dual to A^j. Given functions $f_j = \tilde{f}_j(\vec{\tau})$, define a symmetric $(0,2)$-tensor $h(b)$ and its dual, a self-adjoint $(1,1)$-tensor $h(A)$, by

$$h(b) = \sum_{j=0}^{n-1} f_j b_j, \qquad h(A) = \sum_{j=0}^{n-1} f_j A^j. \tag{1.13}$$

The choice of $h(b)$ appears to be natural: the powers b_j are the only $(0,2)$-tensors which can be obtained algebraically from the second fundamental form b, while $\tau_1, \ldots, \tau_n$ (or, equivalently, $\sigma_1, \ldots, \sigma_n$) generate all the scalar invariants of extrinsic geometry. For example, the Newton transformation $T_r(A)$ depends on all A^j ($j \le r$).

1.2.3 Leaf-Wise Divergence of Operators A^k and $T_r(A)$

Let $\{e_i\}$ ($i \le n$) be a local orthonormal frame of $\mathscr{F}$. If S is a $(1, j+1)$-tensor field S on M, the *divergence* $\operatorname{div} S$ is the $(1, j)$-tensor

$$\operatorname{div} S(Y_1, \ldots, Y_j) = \operatorname{div}_{\mathscr{F}} S(Y_1, \ldots, Y_j) + (\nabla_N S)(N, Y_1, \ldots, Y_j),$$

where the *partial divergence* of S (i.e., along $\mathscr{F}$) is a $(1, j)$-tensor

$$\operatorname{div}_{\mathscr{F}} S(Y_1, \ldots, Y_j) = \sum_{i \le n} (\nabla_{e_i} S)(e_i, Y_1, \ldots, Y_j).$$

The covariant derivative of the $(1, j)$-tensor S is the $(1, j+1)$-tensor given by:

$$(\nabla S)(X, Y_1, \ldots, Y_j) = (\nabla_X S)(Y_1, \ldots, Y_j)$$
$$= \nabla_X(S(Y_1, \ldots, Y_j)) - \sum_{i \le j} S(Y_1, \ldots, \nabla_X Y_i, \ldots Y_j).$$

For example, if S is a $(1,1)$-tensor (a linear operator) then $\operatorname{div} S$ is a vector field. Indeed, the divergence $\operatorname{div} X$ of a vector field X is the scalar function $\operatorname{div}(X^\flat)$ on M, where $X^\flat = g(X, \cdot)$ is the 1-form (i.e., the $(0,1)$-tensor) dual to X. (The "musical" isomorphism $\flat : TM \to T^*M$ sends a vector $X = X^i \partial_i$ to $X^\flat = X_i dx^i = g_{ij} X^j dx^i$.)

For any $X, Y \in TM$, define a linear operator $R_{X,Y} : T\mathscr{F} \to T\mathscr{F}$ by

$$R_{X,Y} : V \to (R(V,X)Y)^\top \qquad (V \in T\mathscr{F}), \qquad (1.14)$$

where R is the Riemannian curvature tensor. For short, write $R_N = R_{N,N}$ and call it the *Jacobi operator*.

Lemma 1.4. *The leafwise divergence of A^k, $k > 0$, satisfies the inductive formula*

$$\operatorname{div}_{\mathscr{F}}(A^k) = A(\operatorname{div}_{\mathscr{F}} A^{k-1}) + (1/k)\nabla^{\mathscr{F}} \tau_k - \sum_{i \leq n} R(N, A^{k-1} e_i) e_i^\top. \qquad (1.15)$$

Equivalently,

$$\operatorname{div}_{\mathscr{F}}(A^k) = \sum_{1 \leq j \leq k} \left(\frac{1}{k-j+1} A^{j-1} \nabla^{\mathscr{F}} \tau_{k-j+1} - \sum_{1 \leq i \leq n} A_N^{j-1} R\left(N, A^{k-j} e_i\right) e_i \right). \qquad (1.16)$$

Moreover, for any vector field $X \perp N$, we have for $k > 0$

$$g(\operatorname{div}_{\mathscr{F}} A^k, X) = \sum_{1 \leq j \leq k} \left(\frac{1}{k-j+1} A^{j-1} X(\tau_{k-j+1}) - \operatorname{Tr}_{\mathscr{F}} \left(A^{k-j} R_{A_N^{j-1} X, N} \right) \right). \qquad (1.17)$$

Proof. By the *Codazzi equation*, see $(1.5)_2$, we have

$$(\nabla_X A)Y - (\nabla_Y A)X = -R(X,Y)N. \qquad (1.18)$$

In order to verify (1.16), we decompose $A^k = A \cdot A^{k-1}$ for $k > 1$, and calculate

$$\operatorname{div}_{\mathscr{F}} A^k = \sum_{i \leq n} (\nabla_{e_i} A^k) e_i = A(\operatorname{div}_{\mathscr{F}} A^{k-1}) + \sum_{i \leq n} (\nabla_{e_i} A) A^{k-1} e_i. \qquad (1.19)$$

Using (1.18), integrability of $T\mathscr{F}$ which yields $[e_i, X]^\perp = 0$, and symmetries of the curvature tensor, we find for X tangent to $\mathscr{F}$,

$$\sum_{i \leq n} g((\nabla_{e_i} A)(A^{k-1} e_i), X) = \sum_{i \leq n} g(A^{k-1} e_i, (\nabla_{e_i} A)X)$$

$$= \sum_{i \leq n} \left[g(A^{k-1} e_i, (\nabla_X A) e_i - R(e_i, X)N \right]$$

$$= \operatorname{Tr}(A^{k-1}(\nabla_X A)) - \sum_{i \leq n} g(R(N, A^{k-1} e_i) e_i, X).$$

For $X \perp N$, the equality (1.19) gives us

$$g(\operatorname{div}_{\mathscr{F}} A^k, X) = g(A(\operatorname{div}_{\mathscr{F}} A^{k-1}), X) + \operatorname{Tr}(A^{k-1}(\nabla_X A)) - \sum_{i \leq n} g(R(N, A^{k-1} e_i) e_i, X).$$

The above and the identity

$$\operatorname{Tr}\left(A^{k-1}(\nabla_X A)\right) = \frac{1}{k} X(\tau_k)$$

(for $k > 0$) yield (1.15), see Remark 1.2. By induction, (1.16) follows from (1.15). Finally, we conclude that (1.17) is a consequence of (1.16) and

$$g\left(\sum_{i \leq n} A^{j-1} R(N, A^{k-j} e_i) e_i^\top, X\right) = g\left(\sum_{i \leq n} A^{k-j} R_{A^{j-1}X, N} e_i, e_i\right)$$

$$= \operatorname{Tr}_{\mathscr{F}}(A^{k-j} R_{A^{j-1}X, N}). \qquad \square$$

Notice that by Lemma 1.4 we have

$$\operatorname{div}_{\mathscr{F}} h(A) = \sum_{k=0}^{n-1} \left(A^k \nabla^{\mathscr{F}} f_k + f_k \operatorname{div}_{\mathscr{F}} A^k\right).$$

Using Lemma 1.4 with $f_j = (-1)^j \sigma_{r-j}$ $(j \leq r)$, we deduce the following claim for Newton transformations of A.

Lemma 1.5. *The leafwise divergence of $T_r(A)$ satisfies the inductive formula*

$$\operatorname{div}_{\mathscr{F}} T_r(A) = -A(\operatorname{div}_{\mathscr{F}} T_{r-1}(A)) + \sum_{i \leq n} R(N, T_{r-1}(A)e_i)e_i^\top, \quad r > 0.$$

Equivalently, $\operatorname{div}_{\mathscr{F}} T_r(A)$ for $r > 0$ is given by the formula:

$$\operatorname{div}_{\mathscr{F}} T_r(A) = \sum_{1 \leq j \leq r} \sum_{i \leq n} (-A)^{j-1} R(N, T_{r-j}(A)e_i)e_i^\top. \tag{1.20}$$

Moreover, for any vector field $X \in \Gamma(\mathscr{F})$, we have

$$g(\operatorname{div}_{\mathscr{F}} T_r(A), X) = \sum_{1 \leq j \leq r} \left[\operatorname{Tr}_{\mathscr{F}}\left(T_{r-j}(A)R_{(-A)^{j-1}X, N}\right)\right.$$

$$\left. - g([T_{r-j}(A)\nabla_N N, (-A)^{j-1}X], N)\right]. \tag{1.21}$$

Proof. Using the inductive definition of $T_r(A)$, we have

$$\operatorname{div}_{\mathscr{F}} T_r(A) = \nabla^{\mathscr{F}} \sigma_r - A(\operatorname{div}_{\mathscr{F}} T_{r-1}(A)) - \sum_{i \leq n} (\nabla_{e_i} A) T_{r-1}(A)e_i.$$

Similarly to the proof of Lemma 1.4, using Codazzi's equation (1.18), we obtain:

$$\sum_{i \leq n} g((\nabla_{e_i} A)T_{r-1}(A)e_i, X) = \sum_{i \leq n} g(T_{r-1}(A)e_i, (\nabla_{e_i} A)X)$$

$$= \sum_{i \leq n} \left[g\left(T_{r-1}(A)e_i, (\nabla_X A)e_i - R(e_i, X)N\right)\right]$$

$$= \operatorname{Tr}(T_{r-1}(A)(\nabla_X A)) - \sum_{i \leq n} g(R(N, T_{r-1}(A)e_i)e_i, X).$$

By Remark 1.2, we have $X(\sigma_r) = \mathrm{Tr}\,(T_{r-1}(A)\nabla_X A)$ for any $X \in \Gamma(\mathscr{F})$. Hence, the inductive formula (1.20) holds. Then (1.20) follows directly. Finally, from the above, it follows that

$$g(\mathrm{div}_{\mathscr{F}}\, T_r(A), X) = \sum_{1 \le j \le r} \sum_{i \le n} g((-A)^{j-1} R(N, T_{r-j}(A)e_i)e_i^\top, X)$$

for every vector field $X \in \Gamma(\mathscr{F})$. One can obtain (1.21) from the above, using the operator (1.14) and the equality

$$\sum_{i \le n} g((-A)^{j-1} R(N, T_{r-j}(A)e_i)e_i^\top, X) = \mathrm{Tr}\,_{\mathscr{F}}(T_{r-j}(A)R_{(-A_N)^{j-1}X,N}). \qquad \square$$

1.2.4 Leaf-Wise Divergence of Vector Fields $h(A)Z$ and $T_r(A)Z$

The next lemma is important for Propositions 1.1 and 1.2.

Lemma 1.6. *Let $\{e_i\}$ be a local orthonormal frame of $\mathscr{F}$ around $q \in M$ such that $\nabla_X^{\mathscr{F}} e_i(q) = 0$ $(X \in (TM)_q)$. Then, at the point q, for $Z = \nabla_N N$ we have*

$$g(\nabla_{e_i} Z, e_j) = (A^2)_{ij} + g(R(e_i, N)N, e_j) - (\nabla_N A)_{ij} + g(Z, e_i)g(Z, e_j). \qquad (1.22)$$

Proof. First, observe that

$$-g(Z, \nabla_{e_i} e_j) = g(\nabla_{e_i} Z, e_j) + g(\nabla_{e_i} N, \nabla_N e_j) + g(N, \nabla_{e_i} \nabla_N e_j). \qquad (1.23)$$

We have

$$(\nabla_N A)_{ij} = g(Z, \nabla_{e_i} e_j) + g(N, \nabla_N \nabla_{e_i} e_j).$$

Therefore, we obtain at $q \in M$

$$(A^2)_{ij} + g(R(e_i, N)N, e_j) - (\nabla_N A)_{ij} = (A^2)_{ij} - g(R(e_i, N)e_j, N) + N(\nabla_{e_i} \nabla_N e_j)$$

$$= (A^2)_{ij} - g(Z, \nabla_{e_i} e_j) - g(\nabla_{e_i} \nabla_N e_j, N)$$

$$+ g(\nabla_{[e_i, N]} e_j, N). \qquad (1.24)$$

Using (1.23), conditions at q, and

$$\nabla_{e_i} N = g(\nabla_{e_i} N, e_k)e_k, \quad \nabla_N e_i = g(\nabla_N e_i, N)N, \quad (A^2)_{ij} = g(\nabla_{e_i} N, e_k)g(\nabla_{e_k} e_j, N),$$

we can simplify the RHS in the last line in (1.24) as:

$$g(\nabla_{e_i} Z, e_j) - g(Z, e_i)g(Z, e_j).$$

From the above it follows (1.22). $\qquad \square$

Let $h(A)$ be of the form described earlier.

Proposition 1.1. *Let $\mathscr{F}$ be a foliation with a unit normal N on a Riemannian manifold (M,g). Then*

$$\operatorname{div}_{\mathscr{F}}(h(A)Z) = g(\operatorname{div}_{\mathscr{F}} h(A), Z) + \sum_{k<n}\left(f_k\,\tau_{k+2} - \frac{f_k}{k+1}N(\tau_{k+1}) \right)$$

$$+ g(h(A)Z, Z) + \operatorname{Tr}_{\mathscr{F}}(h(A)R_N), \qquad (1.25)$$

where

$$g(\operatorname{div}_{\mathscr{F}} h(A), Z) = \sum_{k<n}\left(A^k Z(f_k) + f_k \sum_{1\le j\le k}\left(\frac{1}{k-j+1}A^{j-1}Z(\tau_{k-j+1}) \right.\right. \qquad (1.26)$$

$$\left.\left. - \operatorname{Tr}_{\mathscr{F}}\left(A^{k-j}R_{A^{j-1}Z,N} \right) \right) \right).$$

Proof. As N is unit, $\nabla_X N^\perp = 0$ for any $X \in TM$. Compute the divergence of the vector field $h(A)Z$,

$$\operatorname{div}_{\mathscr{F}} h(A)Z = \sum_{i\le n} g(\nabla_{e_i}(h(A)Z), e_i) = g(\operatorname{div}_{\mathscr{F}} h(A), Z) + \sum_{i\le n} g(\nabla_{e_i} Z,\, h(A)e_i). \qquad (1.27)$$

The first term in the right-hand side of (1.27) is

$$g(\operatorname{div}_{\mathscr{F}} h(A),\, Z) = \sum_{k<n}\left(g(\nabla^{\mathscr{F}} f_k,\, A^k Z) + f_k\, g(\operatorname{div}_{\mathscr{F}} A^k,\, Z) \right).$$

From the above, using (1.17), we obtain (1.26). Applying (1.22) of Lemma 1.6, we can compute the second term in the right-hand side of (1.27), $\sum_{i\le n} g(\nabla_{e_i} Z, h(A)e_i)$, as

$$\sum_{i\le n}\left(g(A^2 e_i + R(e_i, N)N - (\nabla_N A)e_i,\, h(A)e_i) + g(Z, e_i)g(Z, h(A)e_i) \right)$$

$$= \sum_{k<n} f_k\,\tau_{k+2} + \operatorname{Tr}_{\mathscr{F}}(h(A)R_N) - \operatorname{Tr}(h(A)(\nabla_N A)) + g(h(A)Z, Z).$$

We transform $\operatorname{Tr}(h(A)(\nabla_N A))$ using the definition $h(A) = \sum f_k A^k$, as

$$\operatorname{Tr}(h(A)(\nabla_N A)) = \sum_{k<n}\frac{f_k}{k+1}\operatorname{Tr}(\nabla_N(A^{k+1})) = \sum_{k<n}\frac{f_k}{k+1}N(\tau_{k+1}),$$

see (1.11) of Remark 1.2. Finally, we have

$$\sum_{i\le n} g(\nabla_{e_i} Z, h(A)e_i) = \sum_{k<n}\left(f_k\,\tau_{k+2} - \frac{f_k}{k+1}N(\tau_{k+1}) \right) + \operatorname{Tr}_{\mathscr{F}}(h(A)R_N) + g(h(A)Z, Z).$$

$\square$

From Proposition 1.1 with $f_j = (-1)^j \sigma_{r-j}$ the following claim for the Newton transformations $T_r(A)$ results. For the convenience of the reader, we prove it directly.

Proposition 1.2. *Let $\mathscr{F}$ be a foliation on (M,g), and $Z = \nabla_N N$. Then*

$$\mathrm{div}_{\mathscr{F}}(T_r(A)Z) = g(\mathrm{div}_{\mathscr{F}} T_r(A), Z) - N(\sigma_{r+1}) + \sigma_1 \sigma_{r+1} - (r+2)\sigma_{r+2}$$
$$+ \mathrm{Tr}\,_{\mathscr{F}}(T_r(A)R_N) + g(T_r(A)Z, Z),$$

where

$$g(\mathrm{div}_{\mathscr{F}} T_r(A), Z) = \sum_{1 \le j \le r} \mathrm{Tr}\,_{\mathscr{F}}(T_{r-j}(A)R_{(-A)^{j-1}Z,N}). \tag{1.28}$$

Proof. Notice that (1.28) is (1.21) with $X = Z$. As $\nabla_X N^\perp = 0$ for all $X \in TM$, we can compute the divergence of $T_r(A)Z$ as follows:

$$\mathrm{div}_{\mathscr{F}} T_r(A)Z = \sum_{i \le n} g(\nabla_{e_i}(T_r(A)Z), e_i) = g(\mathrm{div}_{\mathscr{F}} T_r(A), Z) + \sum_{i \le n} g(\nabla_{e_i} Z, T_r(A)e_i).$$

Using (1.22) of Lemma 1.6, we compute $\sum_{i \le n} g(\nabla_{e_i} Z, T_r(A)e_i)$ as:

$$\sum_{i \le n} \left(g(A^2 e_i + R(e_i, N)N - (\nabla_N A)e_i, T_r(A)e_i) + g(Z, e_i)g(Z, T_r(A)e_i) \right)$$
$$= -\mathrm{Tr}\,_{\mathscr{F}}(T_r(A)(\nabla_N A - A^2 - R_N)) + g(T_r(A)Z, Z).$$

By Lemma 1.3, we can write

$$\mathrm{Tr}\,_{\mathscr{F}}(T_r(A)(\nabla_N A - A^2 - R_N)) = N(\sigma_{r+1}) - \sigma_1 \sigma_{r+1} + (r+2)\sigma_{r+2} - \mathrm{Tr}\,_{\mathscr{F}}(T_r(A)R_N).$$

Finally, we have

$$\sum_{i \le n} g(\nabla_{e_i} Z, T_r(A)e_i) = -N(\sigma_{r+1}) + \sigma_1 \sigma_{r+1} - (r+2)\sigma_{r+2}$$
$$+ \mathrm{Tr}\,_{\mathscr{F}}(T_r(A)R_N) + g(T_r(A)Z, Z). \qquad \square$$

From the above, applying the Divergence Theorem to any compact leaf, we get

Theorem 1.1. *Let $\mathscr{F}$ be a foliation with a unit normal N on a Riemannian manifold (M,g). Then on any compact leaf L we have*

$$\int_L \left(\sum_{k<n}(A^k Z(f_k) + f_k \sum_{1 \le j \le k} \left(\frac{1}{k-j+1} A^{j-1} Z(\tau_{k-j+1}) - \mathrm{Tr}\,_{\mathscr{F}}(A^{k-j}R_{A^{j-1}Z,N}) \right) \right.$$
$$+ f_k \tau_{k+2} - \frac{f_k}{k+1} N(\tau_{k+1}) \right) + g(h(A)Z, Z) + \mathrm{Tr}\,_{\mathscr{F}}(h(A)R_N) \Big) d\,\mathrm{vol}_L = 0,$$

$$\int_L \left(\sum_{1 \le j \le r} \mathrm{Tr}\,_{\mathscr{F}}(T_{r-j}(A)R_{(-A)^{j-1}Z,N}) - N(\sigma_{r+1}) + \sigma_1 \sigma_{r+1} - (r+2)\sigma_{r+2} \right.$$
$$+ \mathrm{Tr}\,_{\mathscr{F}}(T_r(A)R_N) + g(T_r(A)Z, Z) \Big) d\,\mathrm{vol}_L = 0.$$

1.3 Integral Formulae for Codimension-One Foliations

1.3.1 New Integral Formulae

All the Integral formulae here follow directly from the results of Sect. 1.2.4 and the Divergence Theorem. Recall that for $Z = \nabla_N N$, we have

$$\operatorname{div}(h(A)Z) = \operatorname{div}_{\mathscr{F}}(h(A)Z) - g(h(A)Z, Z).$$

For a special choice of $f_k = (-1)^k \sigma_{r-k}$, i.e., $h(A) = T_r(A)$, we certainly have

$$\operatorname{div}(T_r(A)Z) = \operatorname{div}_{\mathscr{F}}(T_r(A)Z) - g(T_r(A)Z, Z).$$

Proposition 1.3. *Let $\mathscr{F}$ be a foliation with unit normal N on a Riemannian manifold (M,g). Then, using $g(\operatorname{div}_{\mathscr{F}} h(A), Z)$ of (1.26), we have*

$$\operatorname{div}\left(h(A)Z + \left(\sum_{k<n}\frac{1}{k+1}f_k\tau_{k+1}\right)N\right) = g(\operatorname{div}_{\mathscr{F}} h(A), Z) + \operatorname{Tr}_{\mathscr{F}}(h(A)R_N)$$

$$+ \sum_{k<n}\left(f_k\tau_{k+2} + \frac{1}{k+1}\tau_{k+1}(N(f_k) - f_k\tau_1)\right).$$

Theorem 1.2. *Let $\mathscr{F}$ be a foliation with a unit normal N on a closed Riemannian manifold (M,g). Then, using $g(\operatorname{div}_{\mathscr{F}} h(A), Z)$ of (1.26), we have*

$$\int_M\left(g(\operatorname{div}_{\mathscr{F}} h(A), Z) + \sum_{k<n}\left(f_k\tau_{k+2} + \frac{\tau_{k+1}}{k+1}(N(f_k) - f_k\tau_1)\right) + \operatorname{Tr}_{\mathscr{F}}(h(A)R_N)\right)d\operatorname{vol} = 0.$$

$$(1.29)$$

Example 1.2. We look at the first members of the series (1.29). Recall the identity [54]

$$\operatorname{div}(Z + \tau_1 N) = \operatorname{Ric}(N,N) + \tau_2 - \tau_1^2. \qquad (1.30)$$

Let $k = 0$. Because $\tau_1^2 - \tau_2 = 2\sigma_2$, the integrand of (1.29) is $-2\sigma_2 + \operatorname{Ric}(N,N)$, this yields the formula (1.3). For $k = 1$ and $h(A) = A$, (1.29) reads:

$$\int_M\left(\tau_3 + \operatorname{Tr}_{\mathscr{F}}(AR_N) - (1/2)\tau_1\tau_2 + Z(\tau_1) - \operatorname{Tr}_{\mathscr{F}}(R_{Z,N})\right)d\operatorname{vol} = 0. \qquad (1.31)$$

Using (1.30) and identities $3\sigma_3 = \tau_3 + \frac{1}{2}\tau_1^3 - \frac{3}{2}\tau_1\tau_2$, and $Z(\tau_1) = \operatorname{div}(\tau_1 Z) - \tau_1\operatorname{div}Z$, we can rewrite (1.31) in the form:

$$\int_M\left(3\sigma_3 - \sigma_1\operatorname{Ric}(N,N) + \operatorname{Tr}_{\mathscr{F}}(AR_N) - \operatorname{Ric}(N,Z)\right)d\operatorname{vol} = 0. \qquad (1.32)$$

Proposition 1.3 and Theorem 1.2 with a special choice of $f_k = (-1)^k\sigma_{r-k}$ read as

Proposition 1.4. *Let $\mathscr{F}$ be a foliation on a Riemannian manifold (M,g). Then, using $g(\mathrm{div}_{\mathscr{F}} T_r(A), Z)$ of (1.28), we have*

$$\mathrm{div}(T_r(A)Z + \sigma_{r+1}N) = g(\mathrm{div}_{\mathscr{F}} T_r(A), Z) - (r+2)\sigma_{r+2} + \mathrm{Tr}_{\mathscr{F}}(T_r(A)R_N).$$

Theorem 1.3. *Let $\mathscr{F}$ be a foliation on a closed Riemannian manifold (M,g). Then, using $g(\mathrm{div}_{\mathscr{F}} T_r(A), Z)$ of (1.28), we have*

$$\int_M \Big(g(\mathrm{div}_{\mathscr{F}} T_r(A), Z) - (r+2)\sigma_{r+2} + \mathrm{Tr}_{\mathscr{F}}(T_r(A)R_N)\Big) d\mathrm{vol} = 0. \qquad (1.33)$$

Example 1.3. For $r = 0$, (1.33) coincides with (1.3), and for $r = 1$, (1.33) reduces to (1.32). In the next section, we will show that formulae (1.33) generalize (1.4).

1.3.2   Some Consequences of Integral Formulae

From Proposition 1.1 with $k = 0$ it follows

Corollary 1.1. *Let $\mathrm{Ric}(N,N) \geq 0$. Then along any compact leaf with the property $N(\tau_1) \leq 0$, we have*

$$A = 0, \quad \mathrm{Ric}(N,N) = 0, \quad Z = 0.$$

So, if $\mathrm{Ric}(N,N) > 0$ then there are no compact leaves with the property $N(\tau_1) \leq 0$.

Proof. Let $h(A) = \widehat{\mathrm{id}}$. From (1.25) for $k = 0$, using $\mathrm{Tr}\, R_N = \mathrm{Ric}(N,N)$, we have

$$\mathrm{div}_{\mathscr{F}} Z = \tau_2 - N(\tau_1) + \mathrm{Ric}(N,N) + g(Z,Z) \geq 0.$$

Along a compact leaf L, by the Divergence theorem, we obtain $\mathrm{Ric}(N,N) = 0$, $Z = 0$, and $\tau_2 = 0$. From the equality $\tau_2 = 0$ we conclude that $A = 0$. Indeed, if $\mathrm{Ric}(N,N) > 0$ somewhere on a compact leaf L, the above leads to a contradiction,

$$0 < \int_L \Big(\tau_2 - N(\tau_1) + \mathrm{Ric}(N,N) + g(Z,Z)\Big) d\mathrm{vol} = 0. \qquad \square$$

Notice that a Riemannian foliation $\mathscr{F}$ on (M,g) has the property that N-curves are geodesics, in other words, we have $Z = \nabla_N N = 0$.

Hence, from Theorem 1.2 (with $h(A) = A^k$) and Theorem 1.3 it follows

Corollary 1.2. *Let $\mathscr{F}$ be a Riemannian foliation on a closed manifold (M,g). Then*

$$\int_M \Big(\tau_{k+2} - \frac{1}{k+1}\tau_1\,\tau_{k+1} + \mathrm{Tr}_{\mathscr{F}}(A^k R_N)\Big) d\mathrm{vol} = 0, \qquad k \geq 0,$$

$$\int_M \Big((r+2)\sigma_{r+2} - \mathrm{Tr}_{\mathscr{F}}(T_r(A)R_N)\Big) d\mathrm{vol} = 0, \qquad r \geq 0.$$

For $k = r = 0$, the above formulae coincide with (1.3), and for $k = r = 2$, read:

$$\int_M \left(\tau_4 - \frac{1}{3}\tau_1\tau_3 + \mathrm{Tr}\,_{\mathscr{F}}(A^2 R_N) \right) d\,\mathrm{vol} = 0, \quad \int_M \left(\sigma_4(N) - \frac{1}{4}\mathrm{Tr}\,_{\mathscr{F}}(T_2(N)R_N) \right) d\,\mathrm{vol} = 0.$$

One may show that (1.33) by itself yields (1.4). Indeed, let the mixed sectional curvature be constant $c \geq 0$ (i.e., $R_N = c\,\widehat{\mathrm{id}}$). Then

$$R(X,N)Y^\top = 0, \quad \text{for arbitrary vectors } X,Y \in T\mathscr{F},$$

and (1.20) implies that $\mathrm{div}_{\mathscr{F}}\,T_r(A) = 0$ for every $r \geq 0$. By (1.33) and the identity $\mathrm{Tr}\,T_r(A) = (n-r)\,\sigma_r$ (see Lemma 1.3), we get

$$I_{\sigma,r+2} = c\,\frac{n-r}{r+2}\,I_{\sigma,r}, \quad \text{where } I_{\sigma,0} = \int_M 1\,d\,\mathrm{vol} = \mathrm{vol}(M), \quad I_{\sigma,1} = 0,$$

see (1.2). From the above (1.4) follows by a simple induction. Similarly,

$$I_{\tau,0} = n\,\mathrm{vol}(M), \quad I_{\tau,2} = -c\,I_{\tau,0}, \quad \text{etc.}$$

Using (1.29), by induction we get

Corollary 1.3. *Let $\mathscr{F}$ be a minimal foliation (i.e., the mean curvature $H = 0$), and let N define a geodesic foliation on a closed manifold (M,g). If $R_N = c\,\widehat{\mathrm{id}}$ (i.e., the mixed sectional curvature is constant), and n is even, then for any $s > 0$,*

$$I_{\tau,2s+1} = 0, \quad I_{\tau,2s} = (-c)^s\,n\,\mathrm{vol}(M). \tag{1.34}$$

When (M,g) is an *Einstein manifold* with $\dim M > 2$,

$$\mathrm{Ric}(X,Y) = nc \cdot g(X,Y) \quad \text{for all } X,Y \text{ and some } c \in \mathbb{R}$$

and $\mathscr{F}$ is umbilical (i.e., $A = \lambda\,\mathrm{id} = (\tau_1/n)\,\mathrm{id}$), we get formulae similar to (1.4).

Corollary 1.4. *Let $\mathscr{F}$ $(\dim\mathscr{F} = n > 1)$, be an umbilical foliation on a closed Einstein manifold (M^{n+1}, g) (with $\mathrm{Ric} = nc \cdot g$ for some $c \in \mathbb{R}$), then we have*

$$I_{\sigma,k} = \begin{cases} c^{\frac{k}{2}}\binom{n/2}{k/2}\mathrm{vol}(M), & n \text{ and } k \text{ even} \\ 0, & \text{either } n \text{ or } k \text{ odd.} \end{cases}$$

Proof. Let N be a unit normal field and $Z = \nabla_N N$. Under our assumptions,

$$\mathrm{Ric}(N,Z) = 0, \quad \mathrm{Ric}(N,N) = nc.$$

In this case, $T_r(A) = \frac{n-r}{n}\,\sigma_r\,\mathrm{id}$. Hence, see (1.28) and (1.33),

$$g(\mathrm{div}_{\mathscr{F}}\,T_r(A),\,Z) = 0, \quad \mathrm{Tr}\,_{\mathscr{F}}(T_r(A)\,R_N) = c\,\frac{n-r}{r}\,\sigma_r.$$

Then (1.33) reads:

$$\int_M \left((r+2)\,\sigma_{r+2} - nc\,\frac{n-r}{n}\,\sigma_r \right) d\,\mathrm{vol} = 0 \quad \Rightarrow \quad I_{\sigma,r+2} = c\,\frac{n-r}{r+2}\,I_{\sigma,r},$$

From the above, the claim follows by induction. $\qquad\qquad\qquad\Box$

1.3.3 Foliations Whose Leaves Have Constant σ_2

The study of hypersurfaces with constant higher-order mean curvatures has been of increasing interest in recent years (see [1, 3, 6]). Now, we apply IF to provide some results for foliations whose leaves have constant σ_2.

Proposition 1.5. *Let (M,g) be a closed Einstein manifold with non-negative sectional curvature and $\mathscr{F}$ a foliation of M whose leaves have constant σ_2. Then σ_2 must be constant on M.*

Proof. By [6, Proposition 2.31], either σ_2 is constant on M (so the assertion of our proposition is satisfied) or there exists a closed leaf L of $\mathscr{F}$ having the property

$$\sigma_{2|L} = \alpha, \tag{1.35}$$

where $\alpha = \max_M \sigma_2$. On the other hand, M has non-negative sectional curvature, thus $\mathrm{Ric}(N,N) \geq 0$, and (1.3) implies that $\int_M \sigma_2 \,d\,\mathrm{vol} \geq 0$. If σ_2 is not constant on M this implies that $\alpha > 0$ and σ_2 is positive on L. Then, $\sigma_1^2 \geq 2\sigma_2 > 0$ and consequently $\sigma_1 \neq 0$ on L. Without loss of generality, we may assume that $\sigma_1 > 0$ on L. As the eigenvalues of $T_1(A)$ are of the form $\sigma_1 - k_i$ and

$$\sigma_1^2 = \sum_{i=1}^n k_i^2 + 2\sigma_2 > k_i^2,$$

we infer that $T_1(A)$ is positive definite on L. This and the assumption of sectional curvature give

$$\mathrm{Tr}\,_{\mathscr{F}}(T_1(A)R_N) \geq 0 \quad \text{and} \quad g(T_1(A)Z,Z) \geq 0.$$

From (1.35) we conclude that the derivative $N(\sigma_2)$ vanishes on L. On the other hand, we also have $\mathrm{Ric}(N,Z) = 0$, so that from Theorem 1.1 with $r = 1$ we get

$$0 = \int_L \left(\mathrm{Tr}\,_{\mathscr{F}}(A^2 T_1(A)) + \mathrm{Tr}\,_{\mathscr{F}}(R_N T_1(A)) + g(T_1(A)Z,Z) \right) d\,\mathrm{vol}_L > 0.$$

Thus, we arrived at a contradiction, which shows σ_2 is constant on M. $\qquad\Box$

Similarly, we get the following.

Proposition 1.6. *Let (M, g) be an Einstein (not necessarily compact) manifold with non-negative sectional curvature. A leaf of a foliation of M, whose leaves have the same positive constant σ_2, cannot be compact.*

Proof. Assume that a foliation with the above-mentioned properties has a closed leaf L. As before, we obtain that $T_1(A)$ is positive definite on M. As the leaves have the same constant σ_2, then $N(\sigma_2) = 0$. Moreover, $\mathrm{Ric}(N, Z) = 0$, thus the Divergence Theorem and the Proposition 1.2 applied to L yield a contradiction.   □

Chapter 2
Variational Formulae

Abstract We study extrinsic geometry (properties depending on the second fundamental form) of a codimension-one foliation subject to ($\mathscr{F}$-truncated) variations of metrics along the leaves. In Sect. 2.3.1 we develop formulae for the deformation of geometric quantities as the Riemannian metric varies along the leaves. Then, in Sect. 2.3.2, we study variation properties of the functionals depending on the principal curvatures of the leaves and the $\mathscr{F}$-truncated families of metrics, in particular, for conformal metrics along the leaves. The last section of Chap. 2 contains applications to umbilical foliations and minimization of the total bending of the unit normal vector field.

2.1 Introduction

The problem of minimizing geometric quantities has been very popular for many years: recall, for example, classical isoperimetric inequalities, Fenchel estimates of total curvature of curves (and some generalizations like these in [29]), and Kuiper's work on tight and taut submanifolds (see [15] and the bibliographies therein).

In the context of foliations, one has several results of Langevin (and co-authors) [26–28] and so on, and the authors' work [47]. In all cases mentioned earlier, one considers a fixed Riemannian manifold and looks for geometric objects (curves, hypersurfaces, foliations) minimizing geometric quantities defined usually as integrals of curvatures of different types. On the other hand, there is some interest ([33,34,48,50], and so on) in prescribing geometric quantities of given objects (say, foliations): given a foliated manifold $(M, \mathscr{F})$ and a geometric quantity Q (function, vector or tensor field) one may search for a Riemannian metric g on M for which a given geometric invariant (say, curvature of some sort) coincides with Q.

In Chap. 2, the authors describe a new approach combining the two we just mentioned: given a foliated manifold $(M, \mathscr{F})$ and a geometric quantity Q (say, integral of a curvature-like invariant) we look for Riemannian metrics which minimize Q in the class of $\mathscr{F}$-truncated metrics (i.e., the unit vector field N

V. Rovenski and P. Walczak, *Topics in Extrinsic Geometry of Codimension-One Foliations*, SpringerBriefs in Mathematics, DOI 10.1007/978-1-4419-9908-5_2,

orthogonal to $\mathcal{F}$ is the same for all metrics of the variation family). Certainly (as in some of the cases mentioned before) such Riemannian structures need not exist, but if they do, they usually have interesting geometric properties.

The key objective of Chap. 2 is to study properties of Riemannian structures minimizing a quantity Q (for $\mathcal{F}$-truncated metrics). Here Q is built of the invariants of extrinsic geometry of $\mathcal{F}$, that is of the second fundamental forms of the leaves and their invariants arising from its characteristic polynomial: symmetric functions of the principal curvatures.

We denote by $\mathcal{M} = \mathcal{M}(M,\mathcal{F},N)$ the space of smooth Riemannian structures of finite volume on M with N being a unit normal to $\mathcal{F}$. Elements of $\mathcal{M}$ will be called $\mathcal{F}$-truncated metrics. Let $\mathcal{M}_1 \subset \mathcal{M}$ be the subspace of metrics of unit volume. By $\mathcal{F}\mathcal{M}$ (and $\mathcal{F}\mathcal{M}_1$), we denote the foliation on $\mathcal{M}$ (respectively, $\mathcal{M}_1$) by leaves consisting of metrics conformally equivalent along $\mathcal{F}$.

Given $f \in C^2(\mathbb{R}^n)$, we study variational properties of the functional

$$I_f : g \mapsto \int_M f(\vec{\tau})\, \mathrm{d\,vol}_g, \quad \vec{\tau} = (\tau_1,\ldots,\tau_n), \quad g \in \mathcal{M} \tag{2.1}$$

and related functionals (total mean curvatures σ_i, power sums τ_i, etc.). In Sect. 3.7, the function f is the trace of a $(1,1)$-tensor.

2.2   Auxiliary Results

2.2.1   Biregular Foliated Coordinates

The coordinate system described in the following lemma, see [15, Sect. 5.1], is here called a *biregular foliated chart*.

Lemma 2.1. *Let $(M,\mathcal{F},N)$ be a differentiable manifold with a codimension-one foliation $\mathcal{F}$ and a vector field N transversal to $\mathcal{F}$. Then for any $q \in M$ there exists a coordinate system $(x_0,x_1,\ldots x_n)$ on a neighborhood $U_q \subset M$ (centered at q) such that the leaves on U_q are given by $\{x_0 = c\}$ (hence the coordinate vector fields $\partial_i = \partial_{x_i}$, $i \geq 1$, are tangent to leaves), and N is directed along $\partial_0 = \partial_{x_0}$ (one may assume $N = \partial_0$ at q).*

If $(M,\mathcal{F},g)$ is a foliated Riemannian manifold and N is the unit normal then in biregular foliated coordinates $(x_0,x_1,\ldots,x_n)$, g has the form:

$$g = g_{00}\, \mathrm{d}x_0^2 + \sum_{i,j>0} g_{ij}\, \mathrm{d}x_i \mathrm{d}x_j. \tag{2.2}$$

As usual, let g^{ij} denote the entries of the matrix inverse to (g_{ij}) and $g_{ij,k}$ be the derivative of g_{ij} in the direction of ∂_k.

Lemma 2.2. *For a metric (2.2) in biregular foliated coordinates (of a codimension-one foliation $\mathcal{F}$) on (M, g), one has*

$$N = \partial_0 / \sqrt{g_{00}} \quad \text{(the unit normal)}$$

$$\Gamma_{i0}^j = (1/2) \sum_s g_{is,0} g^{sj}, \quad \Gamma_{00}^i = -(1/2) \sum_s g_{00,s} g^{si}, \quad \Gamma_{ij}^0 = -g_{ij,0}/(2g_{00}),$$

$$b_{ij} = \Gamma_{ij}^0 \sqrt{g_{00}} = -\frac{1}{2} g_{ij,0} / \sqrt{g_{00}} \quad \text{(the second fundamental form)},$$

$$A_i^j = -\Gamma_{i0}^j / \sqrt{g_{00}} = \frac{-1}{2\sqrt{g_{00}}} \sum_s g_{is,0} g^{sj} \quad \text{(the Weingarten operator)},$$

$$(b_m)_{ij} = A_{s_2}^{s_1} \dots A_{s_m}^{s_{m-1}} A_i^{s_m} g_{js_1} \quad \text{(the mth "power" of } b_{ij}),$$

$$\tau_i = \left(\frac{-1}{2\sqrt{g_{00}}} \right)^i \sum_{\{r_a\},\{s_b\}} g_{r_1 s_2,0} \cdots g_{r_{i-1} s_i,0} g^{s_1 r_1} \dots g^{s_i r_i}.$$

In particular,

$$\tau_1 = \frac{-1}{2\sqrt{g_{00}}} \sum_{r,s} g_{rs,0} g^{rs}, \quad \tau_2 = \frac{1}{4g_{00}} \sum_{r_1, r_2, s_1, s_2} g_{r_1 s_2,0} g_{r_2 s_1,0} g^{s_1 r_1} g^{s_2 r_2}, \quad \text{etc.}$$

Proof. This is standard and left to the reader. For convenience, observe that the formula for A follows from that for b and $A_i^j = \sum_s b_{is} g^{sj}$. Notice that $(A^m)_i^j = \sum_{\{s_l\}} A_{s_2}^j A_{s_3}^{s_2} \dots A_{s_m}^{s_{m-1}} A_i^{s_m}$. Formulae for b_m follow from the above and $(A^m)_i^s g_{sj} = g(A^m e_i, e_j) = (b_m)_{ij}$. Formulae for τ's follow directly from the above and the equality $\tau_i = \text{Tr}(A^i)$.   □

For example, we apply Lemma 2.2 to the tensor $h(b)$ of (1.13).

Proposition 2.1. *The tensor $h(b)$ of (1.13) with generic functions f_0, f_1 in biregular foliated coordinates has a form:*

$$h(b)_{ac} = f_0(0) g_{ac} - \frac{f_{0,\tau_1}(0)}{2\sqrt{g_{00}}} \left(\sum_{i,j} g^{ij} \psi_{ij} \right) g_{ac} + f_1(0) \psi_{ac} + o([\psi_{ac}]), \quad (2.3)$$

where $\psi_{ij} = g_{ij,0}$. If f_2 is also generic and $f_0(0) = f_{0,\tau_1}(0) = f_1(0) = 0$ then the second-order approximation of $h(b)$ is

$$h(b)_{ac} = \left(\frac{f_{0,\tau_2}(0)}{4g_{00}} \sum_{i,r,s_1,s_2} g^{s_1 i} g^{s_2 r} \psi_{rs_1} \psi_{is_2} + \frac{f_{0,\tau_1 \tau_1}(0)}{8g_{00}} \left(\sum_{i,j} g^{ij} \psi_{ij} \right)^2 \right) g_{ac}$$

$$- \frac{f_{1,\tau_1}(0)}{2\sqrt{g_{00}}} \left(\sum_{i,j} g^{ij} \psi_{ij} \right) \psi_{ac} + \frac{f_2(0)}{4g_{00}} \sum_{i,j} g^{ij} \psi_{ai} \psi_{cj} + o([\psi_{ac}]^2). \quad (2.4)$$

Proof. Take biregular foliated coordinates on M (with the origin at $q \in M$) as in Lemma 2.1. By definition (1.13), the $\mathscr{F}$-components of the tensor are

$$h(b)_{ac} = \sum_{m=0}^{n-1} f_m(\overrightarrow{\tau}) g(A^m e_a, e_c).$$

We neglect the third (and more) order terms with $\psi_{ac} = g_{ac,0}$. By Lemma 2.2 we have $(A^2)_i^j = \sum_r A_r^j A_i^r = \frac{1}{4g_{00}} \sum_{s_1,s_2,r} \psi_{rs_1} g^{s_1 j} \psi_{is_2} g^{s_2 r}$ and, in fact, $\tau_i = o([\psi_{ac}]^{i-1})$ for $i > 0$. From above we obtain expansions for f_j ($j = 0, 1, 2$) as:

$$f_0(\overrightarrow{\tau}) = f_0(\tau_1, \tau_2, 0, \ldots, 0) + o([\psi_{ac}]^2) = f_0(0)$$

$$-\frac{1}{2\sqrt{g_{00}}} f_{0,\tau_1}(0) \sum_{i,j} g^{ij} \psi_{ij} + \frac{1}{8g_{00}} f_{0,\tau_1\tau_1}(0) \left(\sum_{i,j} g^{ij} \psi_{ij}\right)^2$$

$$+\frac{1}{4g_{00}} f_{0,\tau_2}(0) \sum_{i,r,s_1,s_2} \psi_{rs_1} \psi_{is_2} g^{s_1 i} g^{s_2 r} + o([\psi_{ac}]^2),$$

$$f_1(\overrightarrow{\tau}) = f_1(0) - \frac{1}{2\sqrt{g_{00}}} f_{1,\tau_1}(0) \sum_{i,j} g^{ij} \psi_{ij} + o([\psi_{ac}]),$$

$$f_2(\overrightarrow{\tau}) = f_2(0) - \frac{1}{2\sqrt{g_{00}}} f_{2,\tau_1}(0) \sum_{i,j} g^{ij} \psi_{ij} + o([\psi_{ac}]).$$

Substituting into the function $h(b)$ with b_m and τ's from Lemma 2.2,

$$h(b)_{ac} = \sum_{m,s_i,r_j} \frac{(-1)^m f_m(\overrightarrow{\tau})}{(2\sqrt{g_{00}})^m} \psi_{ar_1} \psi_{s_1 r_2} \cdots \psi_{s_{m-1}c} g^{s_1 r_1} \cdots g^{s_{m-1} r_{m-1}},$$

we obtain the second-order approximation of $h(b)$, see (2.4),

$$h(b)_{ac} \approx f_0(0) g_{ac} - \frac{f_{0,\tau_1}(0)}{2\sqrt{g_{00}}} \left(\sum_{i,j} g^{ij} \psi_{ij}\right) g_{ac} + f_1(0) \psi_{ac}$$

$$+\left(\frac{f_{0,\tau_2}(0)}{4g_{00}} \sum_{i,r,s_1,s_2} g^{s_1 i} g^{s_2 r} \psi_{rs_1} \psi_{is_2} + \frac{f_{0,\tau_1\tau_1}(0)}{8g_{00}} \left(\sum_{i,j} g^{ij} \psi_{ij}\right)^2\right) g_{ac}$$

$$-\frac{f_{1,\tau_1}(0)}{2\sqrt{g_{00}}} \left(\sum_{i,j} g^{ij} \psi_{ij}\right) \psi_{ac} + \frac{f_2(0)}{4g_{00}} \sum_{i,j} g^{ij} \psi_{ai} \psi_{cj} + o([\psi_{ac}]^2).$$

From above it follows the first-order approximation (2.3) of $h(b)$. $\square$

Example 2.1. (a) If $h = f(\overrightarrow{\tau}) \hat{b}_1$ and $f(0) \neq 0$ then (2.3) has a form:

$$h(b)_{ac} = f(0) \psi_{ac} + o([\psi_{ac}]).$$

(b) The extrinsic Ricci tensor (see (3.87), Sect. 3.9.1) is a second-degree polynomial

$$\mathrm{Ric}^{\mathrm{ex}}(g)_{ac} = -\frac{1}{4\,g_{00}}\sum_{i,j} g^{ij}\,\psi_{ai}\psi_{cj} - \frac{1}{2\sqrt{g_{00}}}\left(\sum_{i,j} g^{ij}\psi_{ij}\right)\psi_{ac}.$$

(c) The components of the Newton transformation $T_r(A)$ are given by $(T_r(A))_i^j = (r!)^{-1}\varepsilon_{i_1\ldots i_r i}^{j_1\ldots j_r j} A_{j_1}^{i_1}\ldots A_{j_r}^{i_r}$. Hence $T_r(b)$ is the rth degree polynomial

$$T_r(b)_{ac} = (r!)^{-1}\sum \varepsilon_{i_1\ldots i_r a}^{j_1\ldots j_r j}\left(\frac{-1}{2\sqrt{g_{00}}}\right)^s g^{i_1,m_1}\ldots g^{i_r,m_r} g_{jc}\,\psi_{m_1,j_1}\ldots\psi_{m_r,j_r}.$$

(d) For a $\mathscr{F}$-conformal metric $g_{ij} = e^{u(x)}\,\delta_{ij}$ $(i,j > 0)$, we approximate $h(b)$ as

$$h(b)_{ac} = f_0(0)\,\delta_{ac} + u_{,0}\left(f_1(0) - n\frac{f_{0,\tau_1}(0)}{2\sqrt{g_{00}}}\right)\delta_{ac} + o(u_{,0}). \qquad (2.5)$$

If f_0, f_1, f_{0,τ_1} are zero at the origin 0, we use the second-order approximation

$$h(b)_{ac} = u_{,0}{}^2\left(\frac{f_2(0)}{4\,g_{00}} - n\frac{f_{1,\tau_1}(0)}{2\sqrt{g_{00}}}\right)\delta_{ac} + o(u_{,0}{}^2). \qquad (2.6)$$

Let $\hat{g}$ and $g^\perp$ be the components of g along the distributions $T\mathscr{F}$ and $T\mathscr{F}^\perp$, respectively. Because $T\mathscr{F}^\perp$ is one-dimensional, $g^\perp$ is determined by the value $g^\perp(N,N) = 1$.

The next lemma is standard. For the convenience of the reader we give its proof.

Lemma 2.3. *Let $(M, g = \hat{g} \oplus g^\perp)$ be a Riemannian manifold with a codimension-one foliation $\mathscr{F}$ and a unit normal N. Define a metric $\bar{g} = (e^{2\phi}\hat{g}) \oplus g^\perp$, where $\varphi \in C^1(M)$. Then the second fundamental forms and the Weingarten operators of $\mathscr{F}$ with respect to $\bar{g}$ and g are related by*

$$\bar{b} = e^{2\phi}(b - N(\varphi)\hat{g}), \qquad \bar{A} = A - N(\varphi)\,\widehat{\mathrm{id}}. \qquad (2.7)$$

Proof. By the formula for Levi-Civita connection and $g(T\mathscr{F},N) = 0$, we get

$$2e^{2\phi}g(\bar{\nabla}_X N, Y) = 2\bar{g}(\bar{\nabla}_X N, Y) = X(\bar{g}(N,Y)) + N(\bar{g}(X,Y)) - Y(\bar{g}(N,X))$$

$$+\bar{g}([X,N],Y) - \bar{g}([N,Y],X) + \bar{g}([Y,X],N)$$

$$= 2e^{2\phi}N(\phi)g(X,Y) + 2e^{2\phi}g(\nabla_X N,Y)$$

for any vector fields $X,Y \in T\mathscr{F}$. Hence

$$g(\bar{\nabla}_X N, Y) = N(\phi)g(X,Y) + g(\nabla_X N, Y).$$

From this and the definition $\bar{A}(X) = -\bar{\nabla}_X N$, it follows (2.7)$_2$.

Using the definition $\bar{b}(X,Y) = \bar{g}(\bar{\nabla}_X Y, N)$, we compute the tensor $\bar{b}$ on $T\mathscr{F}$

$$\bar{b}(X,Y) = \bar{g}(\bar{\nabla}_X Y, N) = e^{2\phi} g(A(X) - N(\phi)X, Y)$$
$$= e^{2\phi} b(X,Y) - e^{2\phi} g(X,Y) N(\phi) = e^{2\phi} (b(X,Y) - N(\phi) g(X,Y)).$$

From this it follows $(2.7)_1$.   □

Remark 2.1. If $N(\varphi) = 0$ (e.g., φ is constant) then by Lemma 2.3 we have

  (i) $\bar{b} = e^{2\varphi} b$;
 (ii) $\bar{A} = A$;
(iii) $\bar{\tau}_j = \tau_j$ (the power sums).

2.2.2   Foliations with a Time-Dependent Metric

Consider a foliation $(M, \mathscr{F})$ with a time-dependent metric g_t such that $S = \partial_t g$ is an $\mathscr{F}$-truncated tensor. We denote the bundle of $\mathscr{F}$-truncated (k,l)-tensors on M by $\hat{\Lambda}_l^k(M)$. The *inner product of tensors* $F, G \in \hat{\Lambda}_l^k(M)$, denoted by $\langle \cdot, \cdot \rangle$ on M, is given by the following sum:

$$\langle F, G \rangle = g^{a_1 b_1} \ldots g^{a_l b_l} g_{c_1 d_1} \cdots g_{c_k d_k} F_{a_1 \ldots a_l}^{c_1 \ldots c_k} G_{b_1 \ldots b_l}^{d_1 \ldots d_k}.$$

Recall that the *musical isomorphism* $\sharp : T^*M \to TM$ sends a covector $\omega = \omega_i dx^i$ to $\omega^\sharp = \omega^i \partial_i = g^{ij} \omega_j \partial_i$, and $\flat : TM \to T^*M$ sends a vector $X = X^i \partial_i$ to $X^\flat = X_i dx^i = g_{ij} X^j dx^i$. In other words, $X^\flat = g(X, \cdot)$. We denote by $B^\sharp$ the $(1,1)$-tensor field on M which is g-dual to a symmetric $(0,2)$-tensor B,

$$B(X,Y) = g(B^\sharp(X), Y) \quad \text{for all vectors } X, Y.$$

For symmetric $(0,2)$-tensors B, S we have

$$\langle B, S \rangle = \mathrm{Tr}(B^\sharp S^\sharp) = \langle B^\sharp, S^\sharp \rangle. \tag{2.8}$$

Indeed, in a local coordinate basis (e_i) of TM we have

$$g_{ik} B^{ij} = B_k^j, \quad g^{kl} S_{lj} = S_j^k, \quad \langle B, S \rangle = B^{ij} S_{ij}.$$

From the above, in view of identity $g_{ik} g^{kj} = \delta_i^j$ (Kronecker's delta), (2.8) follows,

$$\mathrm{Tr}(B^\sharp S^\sharp) = B_k^j S_j^k = g_{ik} B^{ij} g^{kl} S_{lj} = B^{ij} S_{ij}.$$

For example,
$$\langle \hat{g}, \hat{g} \rangle = \operatorname{Tr} \widehat{\operatorname{id}} = n, \quad \langle \hat{g}, S^2 \rangle = \langle S, S \rangle,$$

where S^2 is the symmetric $(0,2)$-tensor dual to $(S^\sharp)^2$. If $S = s\hat{g}$ for some scalar function s then $S^\sharp = s\,\widehat{\operatorname{id}}$ and has the trace $\operatorname{Tr} S^\sharp = ns$.

Lemma 2.4. *For a smooth family (B_t) of symmetric $(0,2)$-tensors on $(M, g_t, \mathscr{F})$ and $S = \partial_t g_t$, where $g_t \in \mathscr{M}(M, \mathscr{F}, N)$, we have*

$$(\partial_t B_t)^\sharp = \partial_t B_t^\sharp + S^\sharp \cdot B_t^\sharp, \qquad \partial_t(\operatorname{Tr}_{g_t} B_t) = \operatorname{Tr}_{g_t}(\partial_t B_t) - \langle B_t, S \rangle_{g_t}, \quad (2.9)$$

$$\partial_t \langle B_t, S \rangle_{g_t} = \langle \partial_t B_t, S \rangle_{g_t} - 2\langle B_t, S^2 \rangle_{g_t} + \langle B_t, \partial_t S \rangle_{g_t}. \qquad (2.10)$$

Proof. Because $\partial_t g_{ij} = S_{ij}$, we have

$$\partial_t g^{ij} = -S^{ij} := -S_{kl} g^{ik} g^{jl}.$$

To establish $(2.9)_1$, we write $(B^\sharp)_i^k = g^{kj} B_{ij}$ for any t and compute

$$(\partial_t B^\sharp)_i^k = \partial_t (g^{kj} B_{ij}) = \partial_t g^{kj} B_{ij} + g^{kj} \partial_t B_{ij} = -S^{kj} B_{ij} + (\partial_t B)_i^{\sharp k}$$
$$= -g^{ik} S_i^j g_{il} B_j^l + (\partial_t B)_i^{\sharp k} = -S_i^j B_j^k + (\partial_t B)_i^{\sharp k}.$$

Notice that $(2.9)_2$ is a consequence of $(2.9)_1$ and the identity $\operatorname{Tr}(\partial_t B_t^\sharp) = \partial_t(\operatorname{Tr} B_t^\sharp)$. Then (2.10) directly follows from the calculation

$$\partial_t(B^{ij} S_{ij}) = (\partial_t B)^{ij} S_{ij} + B^{ij}(\partial_t S)_{ij} = \partial_t(g^{ai} g^{bj} B_{ab}) S_{ij} + B^{ij}(\partial_t S)_{ij}$$
$$= (\partial_t g)^{ai} g^{bj} B_{ab} S_{ij} + g^{ai}(\partial_t g)^{bj} B_{ab} S_{ij} + g^{ai} g^{bj}(\partial_t B)_{ab} S_{ij} + B^{ij}(\partial_t S)_{ij}$$
$$= -B_{ab}(S^{ai} g^{bj} S_{ij} + g^{ai} S^{bj} S_{ij}) + (\partial_t B)^{ij} S_{ij} + B^{ij}(\partial_t S)_{ij}$$
$$= -B_{ab}\left(g^{al} S_l^i S_i^b + g^{bl} S_l^j S_j^a\right) + (\partial_t B)^{ij} S_{ij} + B^{ij}(\partial_t S)_{ij}$$
$$= -2B_{ab}(S^2)^{ab} + (\partial_t B)^{ij} S_{ij} + B^{ij}(\partial_t S)_{ij}. \qquad \square$$

Now let ∇^t denote the Levi-Civita connection on (M, g_t), where $t \in [0, \varepsilon)$. Observe that the difference of two connections is always a tensor, hence $\Pi_t := \partial_t \nabla^t$ is a $(1,2)$-tensor field on (M, g_t). Differentiating with respect to t, the classical formula for the Levi-Civita connection yields the known formula (see Preface)

$$2g_t(\Pi_t(X, Y), Z) = (\nabla_X^t S)(Y, Z) + (\nabla_Y^t S)(X, Z) - (\nabla_Z^t S)(X, Y), \qquad (2.11)$$

where $X, Y, Z \in TM$. If the vector field $Y = Y(t)$ is time-dependent then

$$\partial_t \nabla_X^t Y = \Pi_t(X, Y) + \nabla_X(\partial_t Y). \qquad (2.12)$$

Notice the symmetry of the tensor: $\Pi_t(X,Y) = \Pi_t(Y,X)$. If the tensor S is $\mathscr{F}$-truncated then by (2.11) we have $\Pi_t(N,\cdot) \perp N$.

Let E be the pull-back of the tangent bundle TM under the projection $M \times [0,\varepsilon) \to M$, i.e., the fiber of E over a point (x,t) is given by $E_{x,t} = T_xM$. There is a natural connection $\tilde{\nabla}$ on E, which extends the Levi-Civita connection ∇ on M. In order to define this connection, we need to specify the covariant t-derivative $\tilde{\nabla}_{\partial_t}$. Given any section X of the vector bundle E, we define

$$\tilde{\nabla}_{\partial_t}X = \partial_t X + \frac{1}{2}S^\sharp(X), \quad \text{in particular,} \quad \tilde{\nabla}_{\partial_t}N = 0. \tag{2.13}$$

The connection $\tilde{\nabla}$ is compatible with the natural bundle metric on E, i.e.,

$$(\tilde{\nabla}_{\partial_t}g)(X,Y) = 0, \qquad X,Y \in TM.$$

To show this, we calculate

$$(\tilde{\nabla}_{\partial_t}g)(X,Y) = \partial_t(g(X,Y)) - g(\tilde{\nabla}_{\partial_t}X,Y) - g(X,\tilde{\nabla}_{\partial_t}Y) = (\partial_t g)(X,Y) - S(X,Y) = 0.$$

This connection is not symmetric: in general $\tilde{\nabla}_{\partial_t}\partial_i \neq 0$, while $\tilde{\nabla}_{\partial_i}\partial_t = 0$ always for $i > 0$. However, each submanifold $M \times \{t\}$ is totally geodesic, so computing derivatives of spatial tangent vector fields gives the same result as computing for sections of $T(M \times [0,\varepsilon))$. In particular, the corresponding Weingarten operators $\tilde{A}$ and A satisfy $\tilde{A} = A$. Clearly, the *torsion tensor*

$$\text{Tor}(X,Y) := \tilde{\nabla}_X Y - \tilde{\nabla}_Y X - [X,Y]$$

vanishes if both arguments are spatial, so the only nonzero components of Tor are

$$\text{Tor}(\partial_t,\partial_i) = \tilde{\nabla}_{\partial_t}\partial_i - \tilde{\nabla}_{\partial_i}\partial_t = \frac{1}{2}S^\sharp(\partial_i), \qquad i > 0.$$

2.2.3 A Differential Operator

To shorten later formulas, we introduce the differential operator

$$\mathscr{V}(F) := \tau_1 F - \nabla_N F,$$

where $F \in \hat{\Lambda}^k_l(M)$ is either a smooth (k,l)-tensor or a function on M.

Lemma 2.5. *For a foliation $\mathscr{F}$ on a closed Riemannian manifold (M,g) we have*

$$\int_M \mathscr{V}(F)\,d\text{vol} = 0 \quad \text{for any} \quad F \in C^1(M).$$

Proof. Using the equality $\operatorname{div} N = -\tau_1$, we have

$$\operatorname{div}(FN) = F \operatorname{div} N + N(F) = -\tau_1 F + N(F) = -\mathscr{V}(F).$$

By the Divergence Theorem, $\int_M \operatorname{div}(FN) \, d\operatorname{vol} = 0$. From the above the required equality follows. $\qquad\square$

The next lemmas (concerning the above operator $\mathscr{V}$) are also global and can be proved for arbitrary (k,l)-tensor fields. First, we show that the operator ∇_N is conjugate to $\mathscr{V}$. Notice that ∇_N commutes with traces of $\mathscr{F}$-truncated $(1,1)$-tensors:

$$\nabla_N(\operatorname{Tr} F) = \operatorname{Tr}(\nabla_N F) \quad \text{for any } (1,1)\text{-tensor field } F.$$

Lemma 2.6. *Let S, B be $\mathscr{F}$-truncated symmetric $(0,2)$-tensor fields on a closed M. Then*

$$\int_M \langle B, \nabla_N S \rangle \, d\operatorname{vol} = \int_M \langle \mathscr{V}(B), S \rangle \, d\operatorname{vol}. \tag{2.14}$$

In particular, for $F \in C^1(M)$ and $S = s\hat{g}$ we have

$$\int_M F N(s) \, d\operatorname{vol} = \int_M s \mathscr{V}(F) \, d\operatorname{vol}. \tag{2.15}$$

Proof. Notice that the $(1,1)$-tensor $\nabla_N S^\sharp$ is g-dual to $\nabla_N S$. To show (2.14), calculate with the help of (2.8) and Lemma 2.5,

$$\int_M \langle B, \nabla_N S \rangle \, d\operatorname{vol} = \int_M \operatorname{Tr}(B^\sharp \nabla_N S^\sharp) \, d\operatorname{vol} = \int_M \left(N(\operatorname{Tr}(B^\sharp S^\sharp)) - \operatorname{Tr}((\nabla_N B^\sharp) S^\sharp) \right) d\operatorname{vol}$$

$$= \int_M \operatorname{Tr}(\tau_1 B^\sharp S^\sharp - (\nabla_N B^\sharp) S^\sharp) \, d\operatorname{vol} = \int_M \langle \mathscr{V}(B), S \rangle \, d\operatorname{vol}.$$

Applying (2.14) to $S = s\hat{g}$, we obtain (2.15). $\qquad\square$

Example 2.2. Let F and s be smooth functions on a closed M. One may use Lemma 2.6 to prove the following:

$$\int_M F s N(s) \, d\operatorname{vol} = \frac{1}{2} \int_M \mathscr{V}(F) s^2 \, d\operatorname{vol}, \tag{2.16}$$

$$\int_M F s N(N(s)) \, d\operatorname{vol} = \int_M \left(\frac{1}{2} \mathscr{V}(\mathscr{V}(F)) s^2 - F N(s)^2 \right) d\operatorname{vol}. \tag{2.17}$$

Indeed, for any $\mathscr{F}$-truncated symmetric $(0,2)$-tensor field S on M, we have

$$\int_M F \langle S, \nabla_N S \rangle \, d\operatorname{vol} = \int_M \langle \mathscr{V}(FS), S \rangle \, d\operatorname{vol} = \int_M \langle \mathscr{V}(F) S, S \rangle - F \langle S, \nabla_N S \rangle \, d\operatorname{vol}.$$

From the above for $S = s\hat{g}$, one obtains (2.16). Using (2.15) with substitution $F \to sF$ and $s \to N(s)$, one has (2.17).

Given linear operator $\Phi : \widehat{\Lambda}_2^0(M) \to \widehat{\Lambda}_2^0(M)$, define

$$\mu(\Phi) := \inf_{S \in \widehat{\Lambda}_2^0(M)} \int_M \langle \Phi(\nabla_N S), \nabla_N S \rangle \, d \operatorname{vol} / \int_M \langle S, S \rangle \, d \operatorname{vol}.$$

Lemma 2.7. *Let a be supremum of the lengths of N-curves.*

(i) *If $a = \infty$ and $\langle \Phi(\nabla_N S), \nabla_N S \rangle \geq 0$ for any $S \in \widehat{\Lambda}_2^0(M)$ then $\mu(\Phi) = 0$.*

(ii) *If $0 \leq \langle \Phi(\nabla_N S), \nabla_N S \rangle \leq b^2 \langle S, S \rangle$ for any $S \in \widehat{\Lambda}_2^0(M)$ and some $b \geq 0$ then*

$$\mu(\Phi) \in [0, \pi^2 b^2 / a^2].$$

Proof. Consider the well-known constrained variation problem

$$J(s) = \int_0^a (s'(x))^2 \, dx \to \min, \quad s(0) = s(a) = 0, \quad \int_0^a (s(x))^2 \, dx = 1,$$

where s is smooth on $(0, a)$. We claim that the minimum of J is π^2 / a^2. Indeed, the *Euler equation* for the functional $\tilde{J}(s) = \int_0^a \tilde{F} \, dx$ with $\tilde{F} = (s')^2 + \mu(s^2 - 1)$ and $\mu \in \mathbb{R}$ is $s'' + \mu s = 0$. Using the boundary conditions at $x \in \{0, a\}$, we find

$$\mu = (\pi/a)^2, \qquad s = C \sin(\pi x / a).$$

From the constraint we calculate $C^2 = 2/a$. The solution (to the constrained variation problem) is

$$\bar{s} = \sqrt{2/a} \sin(\pi x / a), \qquad \min J = J(\bar{s}) = \pi^2 / a^2.$$

Assuming $\bar{s} = 0$ on $\{x < 0\} \cup \{x > 1\}$, we build a piece-wise smooth function $\bar{s}$ on $\mathbb{R}$. Then, there exist functions $s_n \in C^1(\mathbb{R})$ vanishing outside of $(-\frac{1}{n}, a + \frac{1}{n})$ such that $s_n \to \bar{s}$ and $\int_{\mathbb{R}} (s'(x))^2 \, dx \to \pi^2 / a^2$ when $n \to \infty$, the claim is proved.

Let the lengths of N-curves be unbounded and $\langle \Phi(\nabla_N S), \nabla_N S \rangle \geq 0$. Then for any $a > 0$ there exist an N-curve $\gamma : [0, a] \to M$ of length a and a smooth function $s_1 \geq 0$ on γ with the properties

$$s_1(\gamma(0)) = s_1(\gamma(a)) = 0, \qquad \int_\gamma N(s_1)^2 < 2\pi^2 / a^2.$$

There is a "thin" biregular foliated chart $U(q) \supset \gamma$ with $q = \gamma(0)$, $x = (x_0, \hat{x}) \in [0, 1]^{n+1}$, and $\int_0^1 g_{00}(x_0, 0) \, dx_0 = a$. Consider a smooth function $s = s_1(x_0) s_2(\hat{x})$ (with $0 \leq s_2 \leq 1$) supported in this chart. Define a tensor field $S = s\tilde{S}$ on M with an arbitrary $\mathscr{F}$-truncated symmetric $(0,2)$-tensor field $\tilde{S}$ satisfying $\nabla_N \tilde{S} = 0$ on γ, and let $F = \sup_M \langle \Phi(\tilde{S}), \tilde{S} \rangle$. As the volume form vol satisfies $\text{vol} < Q(a) \, dx_0 \wedge d\hat{x}$ along γ for some $Q(a) > 0$ (depending on γ), one may assume that $\text{vol} < 2Q(a) \, dx_0 \wedge d\hat{x}$ on $U(q) = I \times \hat{U}(q)$. Applying the above inequality and the *Fubini Theorem*, we have

$$\int_{U(q)} \langle \Phi(\nabla_N S), \nabla_N S \rangle \, d\text{vol} < F \int_{U(q)} N(s_1)^2 d\text{vol}$$

$$< 2F \int_{U(q)} Q(a) \partial_0(s_1)^2 / g_{00}(x) \, dx_0 \wedge d\hat{x}$$

$$= 2F \, Q(a) \int_{\hat{U}(q)} \frac{1}{g_{00}(x)} \left(\int_I \partial_0(s_1)^2 \, dx_0 \right) d\hat{x}$$

$$< 4Q(a) \left(\frac{\pi}{a} \right)^2 F \int_{\hat{U}(q)} \frac{1}{g_{00}(x)} \, d\hat{x}.$$

One may take $\hat{U}(q)$ such that $4\pi^2 F Q(a) \int_{\hat{U}(q)} \frac{1}{g_{00}(x)} \, d\hat{x} < 1$ for all x_0. Thus,

$$\int_{U(q)} \langle \Phi(\nabla_N S), \nabla_N S \rangle \, d\text{vol} < 1/a^2.$$

As $a > 0$ is arbitrary, it follows that $\mu(\Phi) = 0$, that completes the proof of (i).

Claim (ii) follows directly from the estimates above. $\qquad\qquad\square$

Remark 2.2. Concerning Lemma 2.7(i), recall [20] that there exist compact manifolds (M^{n+1}, g), $n > 2$, foliated by closed curves whose lengths are unbounded.

Lemma 2.8. *Given linear operators* $\Phi_i : \hat{\Lambda}_2^0(M) \rightarrow \hat{\Lambda}_2^0(M)$ $(i = 1, 2, 3)$, *define*

$$J(S) := \int_M \left(\langle \Phi_1(S), \, S \rangle + \langle \Phi_2(\nabla_N S), \, \nabla_N S \rangle + \langle \Phi_3(S), \, \nabla_N S \rangle \right) d\text{vol}.$$

If $J \geq 0$ *for any symmetric tensor* $S \in \Lambda_2^0(M)$ *then* $\langle \Phi_2(\nabla_N S), \nabla_N S \rangle \geq 0$. *Moreover, if*

$$\langle \Phi_2(\nabla_N S), \nabla_N S \rangle \geq 0, \qquad \langle (\Phi_1 + \mathcal{V} \circ \Phi_3)(S), S \rangle \geq -\mu(\Phi_2) \langle S, S \rangle, \qquad (2.18)$$

for any symmetric tensor $S \in \hat{\Lambda}_2^0(M)$ *then* $J \geq 0$.

Proof. By (2.14), we have

$$\int_M \langle \Phi_3(S), \nabla_N S \rangle \, d\text{vol} = \int_M \langle \mathcal{V}(\Phi_3(S)), S \rangle \, d\text{vol}.$$

Certainly, (2.18) with any symmetric tensor $S \in \hat{\Lambda}_2^0(M)$ are suffices for $J \geq 0$.

Fig. 2.1 A saw-shaped function s_1

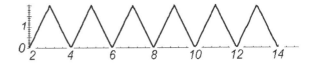

In order to prove that $(2.18)_1$ is necessary, we shall use S with the support in a *biregular foliated chart* $U(q)$ adapted to $\mathscr{F}$ and N with coordinates $x = (x_0, \hat{x}) \in [0,1]^{n+1}$, $\hat{x} = (x_1, \ldots x_n)$, see Sect. 2.2.1. Hence, $x_0 = c = \text{const}$ on the leaves $x_i = c_i = \text{const}$ $(i > 0)$ along N-curves, the coordinate vector fields $\partial_i = \partial_{x_i}$ $(i > 0)$, are tangent to leaves and N is directed along $\partial_0 = \partial_{x_0}$. In these coordinates, the metric g has the form:

$$g = g_{00}\, dx_0^2 + \sum_{i,j>0} g_{ij}\, dx_i dx_j$$

with $g_{00} = 1$ for $\hat{x} = 0$, and $N = \beta\, \partial_0$ for $\beta = 1/\sqrt{g_{00}}$ (because N is the unit normal to $\mathscr{F}$). One may assume that $\text{vol}_g\, U(q) < 1$. Take $s = s_1(x_0) s_2(\hat{x})$, where $0 \le s_i \le 1$ and $\text{supp}(s) \subset U(q)$. Notice that $N(s) = N(s_1)s_2$.

To prove $(2.18)_1$ assume the contrary: that $J \ge 0$ but $F_2 := \langle \Phi_2(\tilde{S}), \tilde{S} \rangle < 0$ for some symmetric $(0,2)$-tensor $\tilde{S}$ at a point q. One may extend $\tilde{S}$ on a neighborhood $U(q)$ (of q) with the property $\nabla_N \tilde{S} = 0$ at q and assume that $F_{2|U(q)} < -\delta$ for some $\delta > 0$. Take a saw-shaped function $s_1 = s_1(x_0)$ (Fig. 2.1) with a number of oscillations of slope ± 1, such that the values of s_1 belong to $[0, \varepsilon]$, where $\varepsilon > 0$ can be chosen as small as necessary. Define a symmetric $(0,2)$-tensor field $S = s\tilde{S}$ on M. Then $N(s_1)^2 = |\partial_0 s_1|^2/g_{00} = 1/g_{00}$ almost everywhere on $[0,1]$, and

$$\int_M \langle \Phi_2(\nabla_N S), \nabla_N S \rangle\, d\,\text{vol} = \int_M \frac{s_2^2}{g_{00}} \langle \Phi_2(\tilde{S}), \tilde{S} \rangle\, d\,\text{vol} < -\delta \int_M \frac{s_2^2}{g_{00}}\, d\,\text{vol} = -\delta\beta_1,$$

where $\beta_1 = \int_M s_2^2/g_{00}\, d\,\text{vol} > 0$. We also have

$$\int_M \left| \langle \Phi_1(S), S \rangle + \langle \Phi_3(S), \nabla_N S \rangle \right| d\,\text{vol} \le F_1 \varepsilon^2 + F_3 \varepsilon,$$

where $F_1 = \sup_{U(q)} |\langle \Phi_1(\tilde{S}), \tilde{S} \rangle|$, $F_3 = \sup_{U(q)} |\langle \Phi_3(\tilde{S}), \tilde{S} \rangle|$. Hence,

$$J(S) < F_1 \varepsilon^2 + F_3 \varepsilon - \beta_1 \delta.$$

For $\varepsilon > 0$ small enough we obtain $J(S) < 0$, a contradiction. $\qquad\qquad\square$

Given the function F on M, define

$$\mu(F) := \inf_{s \in C^1(M)} \int_M F N(s)^2\, d\,\text{vol} \, \Big/ \int_M s^2\, d\,\text{vol}\,.$$

Remark that (by Lemma 2.7) we have the following:

(a) If $a = \infty$ and $F \geq 0$, then $\mu(F) = 0$.

(b) If $0 \leq F \leq b^2$ for some $b \geq 0$, then $\mu(F) \in [0, \frac{\pi^2 b^2}{a^2}]$;

Here a is supremum of the lengths of N-curves.

From Lemma 2.8 we obtain the following

Corollary 2.1. *Let* F_i $(i = 1, 2)$ *be continuous functions on* M *and*

$$J(s) := \int_M (F_1 s^2 + F_2 N(s)^2) \, d\text{vol}.$$

If $J(s) \geq 0$ *for any* $s \in C^1(M)$ *then* $F_2 \geq 0$.

Moreover, if $F_2 \geq 0$ *and* $F_1 \geq -\mu(F_2)$ *then* $J \geq 0$.

2.3 Variational Formulae for Codimension-One Foliations

2.3.1 Variations of Extrinsic Geometric Quantities

In order to calculate variations of the functional I_f with respect to metrics $g_t \in \mathcal{M}$, we find the variational formula for A, and apply it to the Newton transformations $T_i(A)$ and to symmetric functions τ_j, σ_j of A. For short, we shall omit the index t for the time-dependent tensors S, A, $\hat{b}_j$ and functions τ_i, σ_i.

Let $\hat{b}$ be the extension of b to the $\mathcal{F}$-truncated symmetric $(0,2)$-tensor field on M. Notice that $\hat{b}(N, \cdot) = 0$ and

$$\hat{b}(X, Y) = g(A(X), Y).$$

In other words, $\hat{b}(N, \cdot) = 0$ and $\hat{b}$ is dual to the extended Weingarten operator A. Denote by $\hat{b}_j$ the symmetric $(0,2)$-tensor fields on M dual to powers A^j of extended Weingarten operator,

$$\hat{b}_0(X, Y) = \hat{g}(X, Y), \quad \hat{b}_j(X, Y) = \hat{g}(A^j(X), Y) \quad (j > 0, \quad X, Y \in TM).$$

Lemma 2.9. *Let* $g_t \in \mathcal{M}$ *be a family of* $\mathcal{F}$-*truncated metrics and* $S = \partial_t g_t$. *Then the Weingarten operator* A *of* $\mathcal{F}$ *and the symmetric functions* τ_i *and* σ_i *of* A *evolve by*

$$\partial_t A = \frac{1}{2} \left([A, S^\sharp] - \nabla_N^t S^\sharp \right), \tag{2.19}$$

$$\partial_t \tau_i = -\frac{i}{2} \text{Tr}(A^{i-1} \nabla_N^t S^\sharp), \quad \partial_t \sigma_i = -\frac{1}{2} \text{Tr}(T_{i-1}(A) \nabla_N^t S^\sharp), \quad i > 0. \tag{2.20}$$

For $S = s\hat{g}$ $(s \in C^1(M))$ we get

$$\partial_t A = -\frac{1}{2} N(s) \,\hat{\mathrm{id}},$$

$$\partial_t \tau_i = -\frac{i}{2}\, \tau_{i-1} N(s), \quad \partial_t \sigma_i = -\frac{1}{2}(n-i+1)\, \sigma_{i-1} N(s), \quad i > 0.$$

Proof. Using (2.11) and $S(\cdot, N) = 0$ for $\mathscr{F}$-truncated tensors, we obtain

$$\partial_t b(X,Y) = \partial_t g_t(\nabla_X^t Y, N) = (\partial_t g_t)(\nabla_X^t Y, N) + g_t(\partial_t \nabla_X^t Y, N)$$

$$= S(\nabla_X^t Y, N) + (1/2)\left((\nabla_X^t S)(Y,N) + (\nabla_Y^t S)(X,N) - (\nabla_N^t S)(X,Y)\right)$$

$$= (1/2)\left(S(AX,Y) + S(AY,X) - (\nabla_N^t S)(X,Y)\right)$$

for all $X, Y \in T\mathscr{F}$.

Because $S(AX,Y) = g_t(S^\sharp AX, Y)$ and $(\nabla_N^t S)(X,Y) = g_t((\nabla_N^t S^\sharp)X, Y)$, we have

$$g_t((\partial_t A)X, Y) = g_t(\partial_t(AX), Y) = \partial_t b(X,Y) - S(AX,Y)$$

$$= \frac{1}{2}[g_t(S^\sharp AY, X) - g_t(S^\sharp AX, Y) - (\nabla_N^t S)(X,Y)]$$

$$= \frac{1}{2}[g_t([A, S^\sharp]X, Y) - g_t((\nabla_N^t S^\sharp)X, Y)].$$

Formula (2.19) follows from the above and the freedom of choice of $X, Y \in T\mathscr{F}$.
Multiplying (2.19) from the left by A^{i-1}, we get

$$2A^{i-1}\partial_t A = A^{i-1}[A, S^\sharp] - A^{i-1}\nabla_N^t S^\sharp, \quad i > 0.$$

Notice that $\mathrm{Tr}\,(A^{i-1}\cdot[A, S^\sharp]) = 0$, see Remark 2.3. Then, using the identity

$$i\,\mathrm{Tr}\,(A^{i-1}\partial_t A) = \mathrm{Tr}\,(\partial_t A^i) = \partial_t \tau_i,$$

see (1.11) of the following Remark 1.2, we deduce (2.20)$_1$.
Substituting $\partial_t A$ from (2.19) into the formula (1.12) of Remark 1.2, we obtain

$$\partial_t \sigma_i = \frac{1}{2}\mathrm{Tr}\left(T_{i-1}(A)([A, S^\sharp] - \nabla_N^t S^\sharp)\right) = -\frac{1}{2}\mathrm{Tr}\,(T_{i-1}(A)\nabla_N^t S^\sharp),$$

that proves (2.20)$_2$. For $S = s\hat{g}$, we have, respectively, $\nabla_N^t S^\sharp = N(s)\,\hat{\mathrm{id}}$, and

$$\mathrm{Tr}\,(A^{i-1}\nabla_N^t S^\sharp) = \tau_{i-1}N(s), \qquad \mathrm{Tr}\,(T_{i-1}(A)\nabla_N^t S^\sharp) = (n-i+1)\,\sigma_{i-1}N(s).$$

From the above, the case $S = s\hat{g}$ of lemma follows. $\square$

Remark 2.3. For any n-by-n matrices $A, B,$ and C such that $AB = BA$ one has

$$\text{Tr}\,(A \cdot [B,C]) = \text{Tr}\,(ABC) - \text{Tr}\,((AC)B) = \text{Tr}\,(BAC) - \text{Tr}\,(B(AC)) = 0.$$

Example 2.3. (a) We shall find the evolution of tensors A^i and $\nabla_N^t A^i$ and their dual. Using (2.19) and the definition $S = \partial_t g_t$, we generalize (2.19) and find for $i > 0$

$$2\,\partial_t A^i = \sum_{j=0}^{i-1} A^j([A, S^\sharp] - \nabla_N^t S^\sharp) A^{i-1-j} = [A^i, S^\sharp] - \sum_{j=0}^{i-1} A^j (\nabla_N^t S^\sharp) A^{i-1-j}.$$

As $A^i = (\hat{b}_i)^\sharp$, from the above and (2.9) we obtain the evolution of $\hat{b}_i$ $(i > 0)$,

$$(\partial_t \hat{b}_i)^\sharp = \partial_t A^i + S^\sharp A^i = \frac{1}{2}\left(S^\sharp A^i + A^i S^\sharp - \sum_{j=0}^{i-1} A^j (\nabla_N^t S^\sharp) A^{i-1-j}\right).$$

Observe that tracing $\partial_t A^i$ we get $(2.20)_1$.

From (2.19), using (2.11), (2.12), and the following calculations:

$$\begin{aligned}
2\,g_t(\partial_t(\nabla_N^t A^i)X, Y) &= 2\,g_t(\partial_t(\nabla_N^t(A^iX) - A^i\nabla_N^t X), Y) \\
&= 2\,g_t(\partial_t \nabla_N^t(A^iX) - (\partial_t A^i)\nabla_N^t X - A^i\partial_t(\nabla_N^t X), Y) \\
&= 2\,g_t(\nabla_N^t((\partial_t A^i)X) - (\partial_t A^i)\nabla_N^t X, Y) + (\nabla_N^t S)(A^iX, Y) \\
&\quad + (\nabla_{A^iX}^t S)(N, Y) - (\nabla_Y^t S)(N, A^iX) - (\nabla_N^t S)(X, A^iY) \\
&\quad - (\nabla_X^t S)(N, A^iY) + (\nabla_{A^iY}^t S)(N, X) \\
&= 2\,g_t((\nabla_N^t(\partial_t A^i))X, Y) - g_t([[A^i, S^\sharp], A]X, Y) \\
&\quad + g_t([(\nabla_N^t S^\sharp), A^i]X, Y),
\end{aligned}$$

we obtain

$$2\,\partial_t(\nabla_N^t A^i) = 2\,\nabla_N^t(\partial_t A^i) - [[A^i, S^\sharp], A] + [\nabla_N^t S^\sharp, A^i].$$

From the above and the formula for $\partial_t A^i$ we find the evolution of $\nabla_N^t A^i$ $(i > 0)$:

$$\partial_t(\nabla_N^t A^i) = \frac{1}{2}\left([\nabla_N^t A^i, S^\sharp] - [[A^i, S^\sharp], A] - \nabla_N^t \sum_{j=0}^{i-1} A^j (\nabla_N^t S^\sharp) A^{i-1-j}\right).$$

Notice that

$$\partial_t(\nabla_N^t \hat{b}_i)(X, Y) = \partial_t(g_t(\nabla_N^t A^i)X, Y) = S((\nabla_N^t A^i)X, Y) + g_t(\partial_t(\nabla_N^t A^i)X, Y).$$

Putting these facts together yield the evolution equation for $\nabla_N^t \hat{b}_i$ $(i > 0)$:

$$(\partial_t \nabla_N^t \hat{b}_i)^\sharp = \frac{1}{2}\left(S^\sharp \nabla_N^t A^i + (\nabla_N^t A^i)S^\sharp - [[A^i, S^\sharp], A] - \nabla_N^t \sum_{j=0}^{i-1} A^j (\nabla_N^t S^\sharp) A^{i-1-j}\right).$$

(b) Next, we shall find the evolution of tensors $T_i(A)$ and $\nabla_N^t T_i(A)$ and their duals. By Lemma 2.9 and the method of Example 2.3(a) we find the evolution of $T_i(A)$,

$$2\,\partial_t T_i(A) = [T_i(A), S^{\sharp}] - \sum_{j=1}^{i}(-1)^j \sigma_{i-j}\sum_{p=0}^{j-1}A^p(\nabla_N^t S^{\sharp})A^{j-1-p}$$
$$- \sum_{j=0}^{i-1}(-1)^j \operatorname{Tr}\left(T_{i-j-1}(A)\nabla_N^t S^{\sharp}\right)A^j.$$

For $S = s\hat{g}$ $(s \in C^1(M))$ and $i > 0$ we certainly have

$$\partial_t A^i = -\frac{i}{2}A^{i-1}N(s), \qquad \partial_t T_i(A) = \frac{1}{2}N(s)(n-i)\sum_{j=1}^{i}(-1)^j \sigma_{i-j}A^{j-1}.$$

Notice that $\det A \neq 0$ provides

$$\sum_{j=1}^{i}(-1)^j \sigma_{i-j}A^{j-1} = (T_i(A) - \sigma_i \widehat{\operatorname{id}})A^{-1}.$$

Similar to the result for $\nabla_N^t A^i$ in (a), we find the evolution of $\nabla_N^t T_i(A)$,

$$\partial_t(\nabla_N^t T_i(A)) = \nabla_N^t(\partial_t T_i(A)) + (1/2)\left([\nabla_N^t S^{\sharp}, T_i(A)] - [[T_i(A), S^{\sharp}],A]\right)$$
$$= -\frac{1}{2}\nabla_N^t\left(\sum_{j=1}^{i}(-1)^j \sigma_{i-j}\sum_{p=0}^{j-1}A^p(\nabla_N^t S^{\sharp})A^{j-1-p}\right.$$
$$\left.+\sum_{j=0}^{i-1}(-1)^j \operatorname{Tr}\left(T_{i-j-1}(A)\nabla_N^t S^{\sharp}\right)A^j\right)$$
$$+\frac{1}{2}\left([\nabla_N^t T_i(A), S^{\sharp}] - [[T_i(A), S^{\sharp}],A]\right).$$

As $T_i(A) = (T_i(b))^{\sharp}$, $\nabla_N^t T_i(A) = (\nabla_N^t T_i(b))^{\sharp}$, from the above and (2.9) we find the evolution of $T_i(b)$ and $\nabla_N^t T_i(b)$ for $i > 0$,

$$(\partial_t T_i(b))^{\sharp} = \partial_t T_i(A) + S^{\sharp}T_i(A) = \frac{1}{2}\left(S^{\sharp}T_i(A) + T_i(A)S^{\sharp}\right.$$
$$-\sum_{j=1}^{i}(-1)^j \sigma_{i-j}\sum_{p=0}^{j-1}A^p(\nabla_N^t S^{\sharp})A^{j-1-p}$$
$$\left.-\sum_{j=0}^{i-1}(-1)^j \operatorname{Tr}\left(T_{i-j-1}(A)\nabla_N^t S^{\sharp}\right)A^j\right),$$
$$(\partial_t \nabla_N^t T_i(b))^{\sharp} = \partial_t \nabla_N^t T_i(A) + S^{\sharp}\nabla_N^t T_i(A)$$
$$= \frac{1}{2}\left(S^{\sharp}\nabla_N^t T_i(A) + (\nabla_N^t T_i(A))S^{\sharp} - [[T_i(A), S^{\sharp}], A]\right.$$
$$-\nabla_N^t\left(\sum_{j=0}^{i-1}(-1)^j \operatorname{Tr}\left(T_{i-j-1}(A)\nabla_N^t S^{\sharp}\right)A^j\right.$$
$$\left.\left.+\sum_{j=1}^{i}(-1)^j \sigma_{i-j}\sum_{p=0}^{j-1}A^p(\nabla_N^t S^{\sharp})A^{j-1-p}\right)\right).$$

Notice that using the connection (2.13), we also have

$$(\widetilde{\nabla}_{\partial_t} b)(X,Y) = \partial_t b(X,Y) - b(\widetilde{\nabla}_{\partial_t} X, Y) - b(X, \widetilde{\nabla}_{\partial_t} Y) = -\frac{1}{2}(\nabla_N^t S)(X,Y).$$

Proposition 2.2. *Let* $g_t \in \mathcal{M}$ *be a family of* $\mathcal{F}$-*truncated metrics and* $S = \partial_t g_t$. *Then*

$$\widetilde{\nabla}_{\partial_t} \hat{b}_i = -\frac{1}{2} \sum_{j=0}^{i-1} \left(A^j (\nabla_N^t S^\sharp) A^{i-1-j} \right)^\flat,$$

$$\widetilde{\nabla}_{\partial_t} (\nabla_N^t \hat{b}_i) = -\frac{1}{2} \left([[A^i, S^\sharp], A] + \nabla_N^t \sum_{j=0}^{i-1} A^j (\nabla_N^t S^\sharp) A^{i-1-j} \right)^\flat,$$

$$\widetilde{\nabla}_{\partial_t} T_i(b) = -\frac{1}{2} \left(\sum_{j=1}^{i} (-1)^j \sigma_{i-j} \sum_{p=0}^{j-1} A^p (\nabla_N^t S^\sharp) A^{j-1-p} \right.$$

$$\left. + \sum_{j=0}^{i-1} (-1)^j \operatorname{Tr} \left(T_{i-j-1}(A) \nabla_N^t S^\sharp \right) A^j \right)^\flat,$$

$$\widetilde{\nabla}_{\partial_t} (\nabla_N^t T_i(b)) = -\frac{1}{2} \left([[T_i(A), S^\sharp], A] + \nabla_N^t \left(\sum_{j=0}^{i-1} (-1)^j \operatorname{Tr} \left(T_{i-j-1}(A) \nabla_N^t S^\sharp \right) A^j \right. \right.$$

$$\left. \left. + \sum_{j=1}^{i} (-1)^j \sigma_{i-j} \sum_{p=0}^{j-1} A^p (\nabla_N^t S^\sharp) A^{j-1-p} \right) \right)^\flat.$$

Proof. Using Example 2.3(a), (2.13), and equalities

$$\widetilde{\nabla}_{\partial_t} \hat{b}_i(X,Y) = \partial_t \hat{b}_i(X,Y) - \hat{b}_i(\widetilde{\nabla}_{\partial_t} X, Y) - \hat{b}_i(X, \widetilde{\nabla}_{\partial_t} Y),$$

$$\widetilde{\nabla}_{\partial_t} (\nabla_N^t \hat{b}_i)(X,Y) = \partial_t ((\nabla_N^t \hat{b}_i)(X,Y)) - (\nabla_N^t \hat{b}_i)(\widetilde{\nabla}_{\partial_t} X, Y) - (\nabla_N^t \hat{b}_i)(X, \widetilde{\nabla}_{\partial_t} Y),$$

we find $\widetilde{\nabla}_{\partial_t} \hat{b}_i$ and $\widetilde{\nabla}_{\partial_t}(\nabla_N^t \hat{b}_i)$. Similarly, from Example 2.3(b) and the definition (2.13), the formulae for $\widetilde{\nabla}_{\partial_t} T_i(b)$ and $\widetilde{\nabla}_{\partial_t}(\nabla_N^t T_i(b))$ follow.   □

2.3.2 Variations of General Functionals

Here we develop *variational formulae* for the functional $I_f(g)$ of (2.1), restricted to metrics in $\mathcal{M}$ and $\mathcal{M}_1$, respectively. Let

$$\pi : \mathcal{M} \to \mathcal{M}_1, \qquad \pi(g) = \bar{g} = \left(\operatorname{vol}(M,g)^{-2/n} \hat{g} \right) \oplus g^\perp$$

be the $\mathcal{F}$-conformal projection. Metrics $\bar{g}_t = (\phi_t \hat{g}_t) \oplus g_t^\perp$ with dilating factors $\phi_t = \operatorname{vol}(M,g_t)^{-2/n}$, belong to $\mathcal{M}_1$, i.e., $\int_M d\overline{\operatorname{vol}}_t = 1$.

For a family of metrics $g_t \in \mathcal{M}$, we denote $g = g_0$, $S = \partial_t g_t$, and $\dot{S} = \partial_t S$. Recall [53] that the volume form of g_t evolves as:

$$\partial_t(\mathrm{vol}_t) = \frac{1}{2}(\mathrm{Tr}\, S^\sharp)\,\mathrm{vol}_t. \tag{2.21}$$

We certainly have $\mathrm{Tr}\, S^\sharp = \mathrm{Tr}\,_{g_t} S = \langle \hat{g}_t, S \rangle$. For conformal metrics $g_t = e^{\varphi_t} g$, where (φ_t) is a smooth family of continuous (smooth whenever needed) functions on M, we obtain a conformal tensor $S_t = s_t\, \hat{g}_t$, where $s_t = \partial_t \varphi_t$.

Remark 2.4. (a) Let S and $\dot{S}$ be $\mathcal{F}$-truncated symmetric $(0,2)$-tensor fields. If

$$g_t = g + tS + \frac{1}{2}t^2\dot{S}$$

is a "quadratic in t" variation of the metric $g = \hat{g} \oplus g^\perp \in \mathcal{M}$ then the metrics

$$\bar{g}_t = \pi(g_t) = \left(\phi_t\left(\hat{g} + tS + \frac{1}{2}t^2\dot{S}\right)\right) \oplus g^\perp$$

belong to $\mathcal{M}_1$. For a conformal tensor $S = s\hat{g}$ ($s : M \to \mathbb{R}$), the above metrics are

$$g_t = g + \left(ts + \frac{1}{2}t^2\dot{s}\right)\hat{g}, \qquad \bar{g}_t = \pi(g_t) = \left(\phi_t\left(1 + ts + \frac{1}{2}t^2\dot{s}\right)\hat{g}\right) \oplus g^\perp.$$

One may use the above approximations for finding the 1st and second variations of our functionals at $t = 0$ with respect to general families g_t and $\bar{g}_t$, respectively.
(b) Let $\tilde{f} : M \to (0, \infty)$ be a smooth function constant on the leaves of $\mathcal{F}$, and

$$\tilde{g}(X,Y) = \tilde{f}^2 g(X,Y), \quad X,Y \in T\mathcal{F}.$$

If at least one of vectors X, Y is perpendicular to $\mathcal{F}$, we set $\tilde{g}(X,Y) = g(X,Y)$. A foliated Riemannian manifold $(M, \mathcal{F}, \tilde{g})$ is called a *warped foliation* (with a *warping function* $\tilde{f}$), see [59]. They studied from the point of view of the Gromov–Hausdorff convergence. The warped foliation generalizes the Bergers modification of a metric of S^3 along the fibers of the Hopf fibration (called *Berger spheres*). For warped foliations $g_t = ((1 + ct)\hat{g}) \oplus g^\perp$ ($c \in \mathbb{R}$), A and τ_j do not depend on t, and we have

$$\mathrm{vol}_t = (1 + ct)^{n/2}\mathrm{vol}, \quad \phi = (1 + tc)^{-1}, \quad \bar{g}_t = g.$$

Hence, see (2.1),

$$I_f(g_t) = (1 + ct)^{\frac{n}{2}}I_f(g), \quad I'_f(g) = \frac{n}{2}I_f(g).$$

We conclude that *if g is a critical metric for I_f with respect to $\mathscr{F}$-conformal variations g_t then $I_f(g) = 0$.*

In Theorem 2.1 and its corollaries below, we shall find the variations of functionals I_f with respect to metrics $\bar{g}_t \in \mathscr{M}_1$ and $g_t \in \mathscr{M}$. In the g_t-case, the variations/gradients are given by the same formulae as in the first one but with *underlined terms* deleted. This can be explained as follows: under the π_*-projection (from $T\mathscr{M}$ onto $T\mathscr{M}_1$), the gradient of the functional contains additional (underlined) component. So,

$$\nabla I_f(g) = \frac{1}{2}f\hat{g} - \mathscr{V}(B_f) \quad (\text{in } T\mathscr{M}),$$

while its projection onto $T\mathscr{M}_1$, see (2.22), is

$$\bar{\nabla} I_f(g) = \nabla I_f(g) - \frac{1}{2}I_f(g)\hat{g} = \frac{1}{2}(f - I_f(g))\hat{g} - \mathscr{V}(B_f).$$

The scalar product in $T\mathscr{M}$ is given by $\langle\langle \dot{g}_1, \dot{g}_2 \rangle\rangle = \int_M \langle \dot{g}_1, \dot{g}_2 \rangle \, d\,\text{vol}$. By Lemma 2.5, $\int_M \text{Tr}\,\mathscr{V}(B_f^{\sharp})\,d\,\text{vol} = 0$, hence $\langle\langle \bar{\nabla} I_f(g), \hat{g} \rangle\rangle = 0$.

Theorem 2.1. *The gradient of the functional $I_f : \mathscr{M} \to \mathbb{R}$, see (2.1), and its projection via $\pi_* : T\mathscr{M} \to T\mathscr{M}_1$ are given by:*

$$\nabla I_f(g) = \frac{1}{2}(f - I_f(g))\hat{g} - \mathscr{V}(B_f), \tag{2.22}$$

where $B_f = \sum_{i=1}^{n} \frac{i}{2}f_{,\tau_i}\hat{b}_{i-1}$. The $\mathscr{F}\mathscr{M}_1$- and $\mathscr{F}\mathscr{M}$- components of the gradients are

$$\nabla^{\mathscr{F}} I_f(g) = \left(\frac{1}{2}(f - I_f(g)) - \frac{1}{n}\mathscr{V}(\text{Tr}\,B_f^{\sharp}) \right) \hat{g}, \tag{2.23}$$

where $\text{Tr}\,B_f^{\sharp} = \sum_{i=1}^{n} \frac{i}{2}f_{,\tau_i}\tau_{i-1}$.

The second variation of $I_f(\bar{g}_t)$ (when $S = \partial_t g_t$) at a critical metric $g = \bar{g}_0$ and its restriction to the $\mathscr{F}$-conformal variations (i.e., $S = s\hat{g}$, $s : M \to \mathbb{R}$) are given by:

$$I_f''(\bar{g}_t)|_{t=0} = \int_M (\langle \Phi_1(S), S \rangle + \langle \Phi_2(\nabla_N S), \nabla_N S \rangle + \langle \Phi_3(S), \nabla_N S \rangle)\,d\,\text{vol},$$

$$I_f''(\bar{g}_t)|_{t=0} = \int_M \Phi_f N(s)^2 d\,\text{vol}, \tag{2.24}$$

respectively, where

$$\Phi_f = \frac{1}{4}\sum_{i=2}^{n} i(i-1)\,\tau_{i-2}f_{,\tau_i} + \frac{1}{4}\sum_{i,j=1}^{n} ij\,\tau_{i-1}\tau_{j-1}f_{,\tau_i\tau_j},$$

and the linear operators $\Phi_i : \hat{\Lambda}_2^0(M) \to \hat{\Lambda}_2^0(M)$ $(i = 1, 2, 3)$ *are defined by:*

$$\Phi_1(S^\sharp) = \frac{1}{4}(f - I_f(g))(\operatorname{Tr} S^\sharp)\,\hat{\operatorname{id}} - B_f^\sharp[S^\sharp, A], \qquad \Phi_3(S^\sharp) = -(\operatorname{Tr} S^\sharp)B_f^\sharp,$$

$$\Phi_2(S^\sharp) = \sum_{i,j=1}^n \frac{ij}{4} f_{,\tau_i \tau_j} \operatorname{Tr}(A^{i-1}S^\sharp)A^{j-1} + \sum_{i=2}^n \frac{i}{4} f_{,\tau_i} \sum_{j=0}^{i-2} A^j S^\sharp A^{i-2-j}.$$

Proof. As the metrics $\bar{g}_t$ and g_t are $\mathscr{F}$-conformal with constant scale ϕ_t, by Lemma 2.3 we have $\tau_j(\bar{g}_t) = \tau_j(g_t)$ and $\overline{\operatorname{vol}}_t = \phi_t^{n/2} \operatorname{vol}_t$. Differentiating the last equality and using (2.21), we obtain

$$\partial_t \overline{\operatorname{vol}}_t = (\phi^{\frac{n}{2}})' \operatorname{vol}_t + \phi^{\frac{n}{2}} \partial_t \operatorname{vol}_t = \frac{1}{2}\left(\operatorname{Tr} S^\sharp - \int_M (\operatorname{Tr} S^\sharp)\, d\overline{\operatorname{vol}}_t\right) \overline{\operatorname{vol}}_t.$$

Here, we used the fact that $\phi_0 = 1$ and

$$\phi' = -\frac{2}{n} \operatorname{vol}^{-\frac{2}{n}-1} \int_M \partial_t(d\operatorname{vol}_t) = -\frac{1}{n}\phi^{\frac{n}{2}+1} \int_M (\operatorname{Tr} S^\sharp)\, d\operatorname{vol}_t = -\frac{\phi}{n}\int_M (\operatorname{Tr} S^\sharp)\, d\overline{\operatorname{vol}}_t.$$

Differentiating the functional $I_f(\bar{g}_t)$, we obtain

$$I_f'(\bar{g}_t) = \int_M \left(\partial_t f + \frac{1}{2} f \left(\operatorname{Tr} S^\sharp - \int_M (\operatorname{Tr} S^\sharp)\, d\overline{\operatorname{vol}}_t\right)\right) d\overline{\operatorname{vol}}_t$$

$$= \int_M \left(\partial_t f + \frac{1}{2}(f - I_f(\bar{g}_t)) \operatorname{Tr} S^\sharp\right) d\overline{\operatorname{vol}}_t. \qquad (2.25)$$

Now, we simplify (2.25): for $f = f(\vec{\tau})$, by Lemma 2.9, we have

$$\partial_t f = \sum_{i=1}^n f_{,\tau_i} \partial_t \tau_i = -\operatorname{Tr}\left(\nabla_N^t S^\sharp \sum_{i=1}^n \frac{i}{2} f_{,\tau_i} A^{i-1}\right) = -\operatorname{Tr}\left(B_f^\sharp \nabla_N^t S^\sharp\right), \quad (2.26)$$

where $B_f^\sharp = \sum_{i=1}^n \frac{i}{2} f_{,\tau_i} A^{i-1}$ is dual to B_f. From (2.26), by Lemma 2.6, we have

$$\int_M (\partial_t f) d\overline{\operatorname{vol}}_t = \int_M [\operatorname{Tr}((\nabla_N^t B_f^\sharp)S^\sharp) - N(\operatorname{Tr}(B_f^\sharp S^\sharp))]\, d\overline{\operatorname{vol}}_t = \int_M \langle -\mathscr{V}(B_f), S\rangle_t\, d\overline{\operatorname{vol}}_t.$$

Thus, (2.25) yields (2.22):

$$I_f'(\bar{g}_t) = \int_M \left\langle \frac{1}{2}(f - I_f(\bar{g}_t))\hat{g}_t - \mathscr{V}(B_f),\, S \right\rangle_t d\overline{\operatorname{vol}}_t. \qquad (2.27)$$

For a $(0,2)$-tensor $\tilde{B}_t := \frac{1}{2}(f - I_f(\bar{g}_t))\hat{g}_t - \mathscr{V}(B_f)$ in (2.27), by Lemma 2.4 we have

$$\partial_t \langle \tilde{B}_t, S\rangle_{g_t} = \langle \partial_t \tilde{B}_t, S\rangle_{g_t} - 2\langle \tilde{B}_t, S^2\rangle_{g_t} + \langle \tilde{B}_t, \partial_t S\rangle_{g_t},$$

$$\partial_t \tilde{B}_t = \frac{1}{2}(\partial_t f)\hat{g}_t - \frac{1}{2}I_f'(\bar{g}_t)\hat{g}_t + \frac{1}{2}(f - I_f(\bar{g}_t))S - \partial_t \mathscr{V}(B_f).$$

Differentiating (2.27) at a critical metric $\bar{g}_0$ and using $\tilde{B}_0 = 0$, we get

$$I_f''(\bar{g}_t)|_{t=0} = \int_M \left\langle \frac{1}{2}(\partial_t f)\hat{g} + \frac{1}{2}(f - \underline{I_f(g)})S - \partial_t \mathscr{V}(B_f),\ S \right\rangle d\,\text{vol}. \tag{2.28}$$

One may compute $I_f''(\bar{g}_t)|_{t=0}$ explicitly using Lemma 2.9 and (2.26). We have

$$\partial_t \mathscr{V}(B_f) = (\partial_t \tau_1)B_f + \tau_1 \partial_t B_f - \partial_t \nabla_N^t B_f, \qquad \text{where for } t=0$$

$$(\partial_t B_f)^{\sharp} = \sum_i \frac{i}{2}\left(f_{,\tau_i}\partial_t \hat{b}_{i-1} + (\partial_t f_{,\tau_i})\hat{b}_{i-1}\right)^{\sharp} = \frac{1}{2}(S^{\sharp}B_f^{\sharp} + B_f^{\sharp}S^{\sharp})$$

$$\qquad - \sum_i \frac{i}{2}\left(\frac{1}{2}f_{,\tau_i}\sum_{j=0}^{i-2} A^j(\nabla_N S^{\sharp})A^{i-2-j}\right)$$

$$\qquad + \sum_{j=1}^n \frac{j}{2}f_{,\tau_i \tau_j}\,\text{Tr}\left(A^{j-1}(\nabla_N S^{\sharp})\right)A^{i-1}\right),$$

$$(\partial_t(\nabla_N B_f))^{\sharp} = \sum_i \frac{i}{2}\left(N(f_{,\tau_i})\partial_t \hat{b}_{i-1} + f_{,\tau_i}\partial_t \nabla_N \hat{b}_{i-1} + \nabla_N((\partial_t f_{,\tau_i})\hat{b}_{i-1})\right)^{\sharp}$$

$$\qquad = \frac{1}{2}\left(S^{\sharp}\nabla_N B_f^{\sharp} + (\nabla_N B_f^{\sharp})S^{\sharp}\right)$$

$$\qquad - \sum_i \frac{i}{2}\left(\frac{1}{2}\nabla_N\left(f_{,\tau_i}\sum_{j=0}^{i-2} A^j(\nabla_N S^{\sharp})A^{i-2-j}\right)\right.$$

$$\qquad \left. + \nabla_N\left(A^{i-1}\sum_j \frac{j}{2}f_{,\tau_i \tau_j}\,\text{Tr}(A^{j-1}\nabla_N S^{\sharp})\right)\right) - (1/2)[[B_f^{\sharp}, S^{\sharp}], A].$$

Here we used the t-derivatives of $\hat{b}_{i-1}$ and $\nabla_N^t \hat{b}_{i-1}$ of Example 2.3 and the equality

$$\partial_t(f_{,\tau_i}) = -\sum_{j=1}^n \frac{j}{2}f_{,\tau_i \tau_j}\,\text{Tr}(A^{j-1}\nabla_N^t S^{\sharp}).$$

Substituting the above into (2.28), and using the equalities

$$\text{Tr}([[B_f^{\sharp}, S^{\sharp}], A]S^{\sharp}) = 2\,\text{Tr}(B_f^{\sharp}[S^{\sharp}, A]S^{\sharp}) \quad \text{for commuting } B_f^{\sharp} \text{ and } A,$$

$$\mathscr{V}(B_f^{\sharp}) = \frac{1}{2}(f - \underline{I_f(g)})\,\widehat{\text{id}} \quad \text{at a critical metric,}$$

$$\int_M \text{Tr}(\nabla_N S^{\sharp})\,\text{Tr}(B_f^{\sharp}S^{\sharp})d\,\text{vol} = \int_M \langle \mathscr{V}(\text{Tr}(B_f^{\sharp}S^{\sharp}))\hat{g}, S \rangle d\,\text{vol}$$

$$\qquad = \int_M \mathscr{V}(\text{Tr}(B_f^{\sharp}S^{\sharp}))\,\text{Tr}\,S^{\sharp}d\,\text{vol}$$

$$\qquad = \int_M \left(\text{Tr}(\mathscr{V}(B_f^{\sharp})S^{\sharp})\,\text{Tr}\,S^{\sharp} - \text{Tr}(B_f^{\sharp}\nabla_N S^{\sharp})\,\text{Tr}\,S^{\sharp}\right)d\,\text{vol},$$

(the last one is based on Lemma 2.6) we obtain

$$I_f''(\bar{g}_t)_{|t=0} = \int_M \left\{ \frac{1}{4}(f - I_f(g))(\operatorname{Tr} S^\sharp)^2 - \operatorname{Tr}(B_f^\sharp[S^\sharp, A]S^\sharp) - \operatorname{Tr}(B_f^\sharp \nabla_N S^\sharp)\operatorname{Tr} S^\sharp \right.$$

$$+ \tau_1 \sum_i \frac{i}{2}\left(\frac{1}{2}f_{,\tau_i} \sum_{j \leq i-2} \operatorname{Tr}(A^j(\nabla_N S^\sharp)A^{i-2-j}S^\sharp) + \operatorname{Tr}(A^{i-1}S^\sharp) \right.$$

$$\sum_j \frac{j}{2} f_{,\tau_i\tau_j} \operatorname{Tr}(A^{j-1}\nabla_N S^\sharp) \bigg)$$

$$- \sum_i \frac{i}{2}\left(\frac{1}{2}\operatorname{Tr}\left(S^\sharp \nabla_N \left(f_{,\tau_i} \sum_{j=0}^{i-2} A^j(\nabla_N S^\sharp)A^{i-2-j} \right) \right) \right.$$

$$+ \operatorname{Tr}\left(S^\sharp \nabla_N^t \left(A^{i-1}\sum_j \frac{j}{2} f_{,\tau_i\tau_j} \operatorname{Tr}(A^{j-1}\nabla_N S^\sharp) \right) \right) \bigg) \Bigg\} d\operatorname{vol}.$$

Simplifying terms with $f_{,\tau_i\tau_j}$ and sums "$\sum_{j=0}^{i-2}$" (by Lemma 2.5), we finally obtain

$$I_f''(\bar{g}_t)_{|t=0} = \int_M \left\{ \frac{1}{4}(f - I_f(g))(\operatorname{Tr} S^\sharp)^2 - \operatorname{Tr}(B_f^\sharp[S^\sharp, A]S^\sharp) - \operatorname{Tr}(B_f^\sharp \nabla_N S^\sharp)\operatorname{Tr} S^\sharp \right.$$

$$+ \sum_{i=2}^n \sum_{j=0}^{i-2} \frac{i}{4} f_{,\tau_i} \operatorname{Tr}(A^j(\nabla_N S^\sharp)A^{i-2-j}\nabla_N S^\sharp)$$

$$+ \sum_{i,j=1}^n \frac{ij}{4} f_{,\tau_i\tau_j} \operatorname{Tr}(A^{i-1}\nabla_N S^\sharp)\operatorname{Tr}(A^{j-1}\nabla_N S^\sharp) \bigg\} d\operatorname{vol}. \quad (2.29)$$

By (2.29), the integrand of $I_f''(\bar{g}_t)_{|t=0}$ has the form $(2.24)_1$.

Let $S = s\hat{g}$. Although the result follows from the above (the RHS of the formula for Φ_2 reads as $\Phi_f N(s)^2$ and $\int_M \langle (\Phi_1 + \mathscr{V} \circ \Phi_3)(S), S \rangle d\operatorname{vol} = 0$), we shall prove it independently. In this case, $\operatorname{Tr} B_f^\sharp = \sum_{i=1}^n \frac{i}{2} f_{,\tau_i}\tau_{i-1}$, and (2.27) reads:

$$I_f'(\bar{g}_t) = \int_M s\left(\frac{n}{2}(f - I_f(\bar{g}_t)) - \mathscr{V}(\operatorname{Tr} B_f^\sharp) \right) d\overline{\operatorname{vol}}_t$$

$$= \int_M \left\langle \left(\frac{1}{2}(f - I_f(\bar{g}_t)) - \frac{1}{n}\mathscr{V}(\operatorname{Tr} B_f^\sharp) \right) \hat{g}, s\hat{g} \right\rangle d\overline{\operatorname{vol}}_t. \quad (2.30)$$

From (2.30) with $t = 0$ (or, from (2.22)), (2.23) follows. Differentiating (2.30) at a critical metric $g = \bar{g}_0$, or applying $S = s\hat{g}$ to (2.28), we find

$$I_f''(\bar{g}_t)_{|t=0} = \int_M \left(s\frac{n}{2}\partial_t f + s^2 \frac{n}{2}(f - I_f(g)) - s\operatorname{Tr}_g(\partial_t\mathscr{V}(B_f)) \right) d\operatorname{vol}.$$

From the above formula and (2.23) at a critical metric, we have

$$\int_M \left(s\frac{n}{2}\partial_t f + s^2\frac{n}{2}(\underline{f - I_f(\bar{g}_t)}) \right) d\,vol = \int_M \left(\mathscr{V}(\mathrm{Tr}\,B_f^\sharp)s^2 - \frac{n}{2}(\mathrm{Tr}\,B_f^\sharp)sN(s) \right) d\,vol$$

$$= \left(1 - \frac{n}{4}\right)\int_M \mathscr{V}(\mathrm{Tr}\,B_f^\sharp)s^2\,d\,vol.$$

By Lemma 2.9, we have

$$\partial_t(\mathrm{Tr}\,B_f^\sharp)_{|t=0} = \sum_i \frac{i}{2}((\partial_t f,_{\tau_i})\tau_{i-1} + f,_{\tau_i}\partial_t \tau_{i-1}) = -\Phi_f N(s).$$

Using this, $(2.9)_2$ and the identity

$$\partial_t \mathscr{V}(\phi_t) = (\partial_t \tau_1)\phi_t + \mathscr{V}(\partial_t \phi_t), \quad \forall \phi_t \in C^1(M),$$

we calculate

$$\int_M s\mathrm{Tr}\,_g(\partial_t \mathscr{V}(B_f))d\,vol = \int_M s\left(\partial_t(\mathrm{Tr}\,_{g_t}\mathscr{V}(B_f))_{|t=0} + s\mathrm{Tr}\,_g\mathscr{V}(B_f)\right)d\,vol$$

$$= \int_M \left(\mathscr{V}(\mathrm{Tr}\,B_f^\sharp)s^2 + s\partial_t(\mathscr{V}(\mathrm{Tr}\,B_f^\sharp))_{|t=0} \right)d\,vol$$

$$= \int_M \left(\mathscr{V}(\mathrm{Tr}\,B_f^\sharp)s^2 + s\left(\mathscr{V}(-\Phi_f N(s)) \right. \right.$$

$$\left. \left. +(\partial_t \tau_1)\mathrm{Tr}\,B_f^\sharp \right) \right)d\,vol$$

$$= \int_M \left(\mathscr{V}(\mathrm{Tr}\,B_f^\sharp)s^2 - \mathscr{V}(\Phi_f)sN(s) + \Phi_f sN(N(s)) \right.$$

$$\left. -\frac{n}{2}(\mathrm{Tr}\,B_f^\sharp)sN(s) \right)d\,vol$$

$$= \int_M \left(\left(1 - \frac{n}{4}\right)\mathscr{V}(\mathrm{Tr}\,B_f^\sharp)s^2 - \Phi_f N(s)^2 \right)d\,vol.$$

The formula $(2.24)_2$ follows from the above. $\square$

Example 2.4 (Totally geodesic foliations). Let $\mathscr{F}$ be a totally geodesic foliation on (M,g) of unit volume. Then

$$A = 0, \quad \vec{\tau} = 0, \quad B_f = \frac{1}{2}f,_{\tau_1}(0)\hat{g}, \quad \mathscr{V}(B_f) = 0, \quad I_f(g) = f(0).$$

By (2.22) of Theorem 2.1, g is a critical metric for the functional I_f with respect to variations $\bar{g}_t \in \mathscr{M}_1$. We also have $\Phi_f = \frac{n}{2}f,_{\tau_2}(0) + \frac{n^2}{4}f,_{\tau_1\tau_1}(0)$ and

$$\Phi_1(S^\sharp) = 0, \quad \Phi_3(S^\sharp) = -\frac{1}{2}f,_{\tau_1}(0)(\mathrm{Tr}\,S^\sharp)\widehat{\mathrm{id}},$$

$$\Phi_2(S^\sharp) = \frac{1}{4}f,_{\tau_1\tau_1}(0)(\mathrm{Tr}\,S^\sharp)\widehat{\mathrm{id}} + \frac{1}{2}f,_{\tau_2}(0)S^\sharp.$$

Using Lemmas 2.5 and 2.6, we calculate

$$I_f''(\bar{g}_t)|_{t=0} = \int_M \left(\frac{1}{4} f_{,\tau_1\tau_1}(0)(N(\mathrm{Tr}\, S^\sharp))^2 + \frac{1}{2} f_{,\tau_2}(0)\, \mathrm{Tr}\,((\nabla_N S^\sharp)^2) \right) d\,\mathrm{vol}.$$

We conclude that $I_f'' \geq 0$, when $f_{,\tau_1\tau_1}(0) \geq 0$ and $f_{,\tau_2}(0) \geq 0$.

Question. Under what conditions on a smooth function $f(\tau_1,\dots,\tau_n)$ is the form

$$\langle \Phi_2(S), S \rangle = \sum_{i,j=1}^n \frac{ij}{4} f_{,\tau_i\tau_j}\, \mathrm{Tr}\,(A^{i-1}S^\sharp)\, \mathrm{Tr}\,(A^{j-1}S^\sharp) + \sum_{i=2}^n \frac{i}{4} f_{,\tau_i} \sum_{j=0}^{i-2} \mathrm{Tr}\,(A^j S^\sharp A^{i-2-j} S^\sharp)$$

positive definite for all $\mathscr{F}$-truncated symmetric $(0,2)$-tensors S?

Example. For $f = \tau_2$, we have $\langle \Phi_2(S), S \rangle = \frac{1}{2}\langle S, S \rangle$. Notice that for $S = s\,\widehat{\mathrm{id}}$ the condition reads:

$$\sum_{i,j=1}^n ij\, f_{,\tau_i\tau_j}\, \tau_{i-1}\tau_{j-1} + \sum_{i=2}^n i(i-2) f_{,\tau_i}\, \tau_{i-2} > 0.$$

Remark 2.5. Consider the function

$$F = \sum_{i,j=1}^{n-1} \tilde{H}_{ij} \tau_i \tau_j + 2 \sum_{i=1}^{n-1} b_i \tau_i + c, \quad \text{where} \quad c = n^2 f_{,\tau_1\tau_1} + 2n f_{,\tau_2},$$

while the $(n-1)\times(n-1)$ matrix $\tilde{H}$ and $(n-1)$-vector b are given by:

$$\tilde{H}_{ij} = (i+1)(j+1) f_{,\tau_{i+1}\tau_{j+1}} \quad (1 \leq i,j < n),$$

$$b_i = n(i+1) f_{,\tau_1\tau_{i+1}} - n^2 f_{,\tau_1\tau_n} \delta_{i,n-1} + \frac{1}{2}(i+2)(i+1) f_{,\tau_{i+2}} \quad (1 \leq i < n).$$

Critical points of F are solutions $\tilde{\tau} = (\tau_1,\dots,\tau_{n-1})$ to the system $\tilde{H}\tilde{\tau} = -b$. If $\tilde{H}$ is positive definite and $b^T (\tilde{H}^{-1})^T (\tilde{H} - 2\,\mathrm{id}) b > -c$ for all $\tilde{\tau}$, then $F > 0$. Indeed, under above conditions, Φ_f of Theorem 2.1 is strictly positive and for any $g \in \mathscr{M}_1$ the functional I_f, when restricted on $\mathscr{F}$-conformal metrics of unit volume, has at most one critical point.

Proposition 2.3. *Let a metric g on a closed foliated manifold $(M,\mathscr{F})$ be a stable local maximum on the space $\mathscr{M}_1$ for the functional I_f (with a fixed $f \in C^2(\mathbb{R}^n)$). If $\int_M \langle \Phi_2(\nabla_N S), \nabla_N S \rangle d\,\mathrm{vol} \geq 0$ for any $\mathscr{F}$-truncated symmetric $(0,2)$-tensor S, and*

$$\sum_{m=1}^n m f_{,\tau_m} (k_i^{m-1} - k_j^{m-1})(k_i - k_j) \geq 0 \tag{2.31}$$

for any principal curvatures $k_i \neq k_j$, then $\mathscr{F}$ is umbilical.

Proof. One may take $S^\sharp$ with the property $\mathrm{Tr}\, S^\sharp = 0$. Then, by Theorem 2.1,

$$I_f''(g) = \int_M \left(-\mathrm{Tr}\,(B_f^\sharp [S^\sharp, A] S^\sharp) + \langle \Phi_2(\nabla_N S), \nabla_N S \rangle \right) d\,\mathrm{vol}.$$

Let $S^\sharp = (s_{ij})$ in the frame of principal directions (for A). One may show that

$$-\operatorname{Tr}\left(A^m\left[S^\sharp, A\right]S^\sharp\right) = \sum_{i<j} s_{ij}^2\left(k_i^{m-1} - k_j^{m-1}\right)(k_i - k_j)$$

for all $m > 0$. Hence, by the condition (2.31), for $B_f^\sharp = \sum_{m=1}^n f_{,\tau_m}A^{m-1}$ we have

$$-\operatorname{Tr}\left(B_f^\sharp\left[S^\sharp, A\right]S^\sharp\right) \geq 0.$$

and the equality holds only when $k_i = k_j$ for all $i \neq j$. As the second variation $I_f''(g) \leq 0$, from the above we conclude that $\mathscr{F}$ is umbilical. □

Notice that the condition $f_{,\tau_m}\begin{cases} \geq 0, & \text{for } m \text{ even} \\ = 0, & \text{for } m \text{ odd} \end{cases}$ yields the inequality (2.31).

2.3.3 Variations of Particular Functionals

The following functionals on $\mathscr{M}$ for particular cases of f were introduced in (1.1):

$$I_{\tau,k}(g) = \int_M \tau_k \, d\operatorname{vol}_g, \quad I_{\sigma,k}(g) = \int_M \sigma_k \, d\operatorname{vol}_g, \quad k = 1, 2, \ldots.$$

From [41], see (1.2), it is known that $I_{\tau,1} = I_{\sigma,1} = 0$ for any $\mathscr{F}$ and g on a closed M. From Theorem 2.1 with $f = \tau_k$ it follows

Corollary 2.2. *The gradient of the functional $I_{\tau,k} : \mathscr{M} \to \mathbb{R}$ for $k > 1$ and its projection via $\pi_* : T\mathscr{M} \to T\mathscr{M}_1$ are given by:*

$$\nabla I_{\tau,k}(g) = \frac{1}{2}\left(\tau_k - \overline{I_{\tau,k}(g)}\right)\hat{g} - \frac{k}{2}\mathscr{V}(\hat{b}_{k-1}).$$

The $\mathscr{F}\mathscr{M}$- and $\mathscr{F}\mathscr{M}_1$- components of above gradient are given, respectively, by

$$\nabla^{\mathscr{F}} I_{\tau,k}(g) = \frac{1}{2}\left(\tau_k - \overline{I_{\tau,k}(g)} - \frac{k}{n}\mathscr{V}(\tau_{k-1})\right)\hat{g}.$$

The second variation of $I_{\tau,k}$ at a critical metric $g = \bar{g}_0$, and its restriction to the $\mathscr{F}$-conformal variations $S = s\hat{g}$ ($s : M \to \mathbb{R}$) are given by (2.24), where $S = \partial_t \bar{g}_t$, $f = \tau_k$, $\Phi_f = \frac{1}{4}k(k-1)\tau_{k-2}$, and

$$\Phi_1(S^\sharp) = \frac{1}{4}\left(\tau_k - \overline{I_{\tau,k}(g)}\right)(\operatorname{Tr} S^\sharp)\widehat{\operatorname{id}} - \frac{k}{2}A^{k-1}[S^\sharp, A],$$

$$\Phi_2(S^\sharp) = \frac{k}{4}\sum_{j=0}^{k-2}A^j S^\sharp A^{k-2-j}, \qquad \Phi_3(S^\sharp) = -\frac{k}{2}(\operatorname{Tr} S^\sharp)A^{k-1}.$$

Proof. As in the proof of Theorem 2.1 (see (2.25) with $f = \tau_k$), we obtain

$$I'_{\tau,k}(\bar{g}_t) = \int_M \left(\partial_t \tau_k + \frac{1}{2}(\tau_k - I_{\tau,k}(\bar{g}_t)) \operatorname{Tr} S^\sharp \right) d\overline{\operatorname{vol}}_t.$$

By $(2.20)_1$ and Lemma 2.6, we have at $t = 0$

$$\int_M (\partial_t \tau_k) d\overline{\operatorname{vol}} = -\frac{k}{2} \int_M \langle A^{k-1}, \nabla_N S^\sharp \rangle d\overline{\operatorname{vol}} = -\frac{k}{2} \int_M \langle \mathscr{V}(A^{k-1}), S^\sharp \rangle d\overline{\operatorname{vol}}.$$

The above (or Theorem 2.1 with $B_f = \frac{k}{2}\hat{b}_{k-1}$ and $\operatorname{Tr} B_f^\sharp = \frac{k}{2}\tau_{k-1}$) yield the formula for the gradient $\nabla I_{\tau,k}(g)$.

We shall comment about $I''_{\tau,k}(\bar{g}_t)|_{t=0}$. By (2.28), we obtain

$$I''_{\tau,k}(\bar{g}_t)|_{t=0} = \frac{1}{2} \int_M \left\langle (\partial_t \tau_k)\hat{g} + (\tau_k - I_{\tau,k}(g)) S - k \partial_t \mathscr{V}(\hat{b}_{k-1}), S \right\rangle d\operatorname{vol}. \quad (2.32)$$

Observe that by $(2.20)_1$ the first term in (2.32) yields:

$$\frac{1}{2} \int_M \langle (\partial_t \tau_k)\hat{g}, S \rangle d\operatorname{vol} = -\frac{k}{4} \int_M (\operatorname{Tr} S^\sharp) \operatorname{Tr} (A^{k-1}\nabla_N S^\sharp) d\operatorname{vol}.$$

We have $\partial_t \mathscr{V}(\hat{b}_{k-1}) = (\partial_t \tau_1)\hat{b}_{k-1} + \tau_1 \partial_t \hat{b}_{k-1} - \partial_t \nabla_N^t \hat{b}_{k-1}$, where by Example 2.3(a),

$$\int_M \langle \partial_t \hat{b}_{k-1}, S \rangle d\operatorname{vol} = \int_M \left(\operatorname{Tr}(S^\sharp A^{k-1} S^\sharp) \right.$$
$$\left. -\frac{1}{2} \operatorname{Tr} \left(S^\sharp \sum_{j=0}^{k-2} A^j (\nabla_N S^\sharp) A^{k-2-j} \right) \right) d\operatorname{vol},$$

$$\int_M \langle \partial_t (\nabla_N \hat{b}_{k-1}), S \rangle d\operatorname{vol} = \int_M \left(\operatorname{Tr}(S^\sharp (\nabla_N^t A^{k-1}) S^\sharp) - \operatorname{Tr}(A^{k-1}[S^\sharp, A]S^\sharp) \right.$$
$$\left. -\frac{1}{2} \operatorname{Tr} \left(S^\sharp \nabla_N^t \sum_{j=0}^{k-2} A^j (\nabla_N^t S^\sharp) A^{k-2-j} \right) \right) d\operatorname{vol}.$$

Using the above, and the identities

$$\int_M \operatorname{Tr}(\nabla_N S^\sharp) \operatorname{Tr}(S^\sharp A^{k-1}) d\operatorname{vol} = \int_M \left(\operatorname{Tr} \left(S^\sharp \mathscr{V}(A^{k-1}) - A^{k-1} \nabla_N S^\sharp \right) \operatorname{Tr} S^\sharp \right) d\operatorname{vol},$$

$$k\mathscr{V}(A^{k-1}) = (\tau_k - I_{\tau,k}(g))\widehat{\operatorname{id}} \quad \text{(at a critical metric)}$$

(or directly from (2.29) with $f = \tau_k$) we obtain

$$I_f''(\bar{g}_t)_{|t=0} = \int_M \left\{ \frac{1}{4}(\tau_k - I_{\tau,k}(g))(\operatorname{Tr} S^\sharp)^2 - \frac{k}{2}\operatorname{Tr}(A^{k-1}[S^\sharp, A]S^\sharp) \right.$$
$$\left. - \frac{k}{2}\operatorname{Tr}(A^{k-1}\nabla_N S^\sharp)\operatorname{Tr} S^\sharp + \frac{k}{4}\operatorname{Tr}\left((\nabla_N S^\sharp)\sum_{j=0}^{k-2} A^j (\nabla_N S^\sharp)A^{k-2-j}\right) \right\} d\operatorname{vol}.$$

From the above the required formulae for Φ_i $(i = 1, 2, 3)$ follow. ☐

Notice that for $f = \tau_2$ the form $\langle \Phi_2(\nabla_N S^\sharp), \nabla_N S^\sharp \rangle$ is non-negative definite. By Lemma 2.7(i), if the lengths of N-curves are unbounded then $\mu(\Phi_2) = 0$.

In the next consequence of Theorem 2.1 (for $f = \sigma_k$) we represent the operators Φ_k explicitly using Newtonian transformations.

Corollary 2.3. *The gradient of the functional $I_{\sigma,k} : \mathcal{M} \to \mathbb{R}$ for $k > 1$ and its projection via $\pi_* : T\mathcal{M} \to T\mathcal{M}_1$ are given by:*

$$\nabla I_{\sigma,k}(g) = \frac{1}{2}(\sigma_k - I_{\sigma,k}(g))\,\hat{g} - \frac{1}{2}\mathcal{V}(T_{k-1}(b)).$$

The $\mathcal{F}\mathcal{M}$- and $\mathcal{F}\mathcal{M}_1$- components of above gradients are given, respectively, by:

$$\nabla^{\mathcal{F}} I_{\sigma,k}(g) = \frac{1}{2}\left(\sigma_k - I_{\sigma,k}(g) - \frac{n-k+1}{n}\mathcal{V}(\sigma_{k-1})\right)\hat{g}.$$

The second variation of $I_{\sigma,k}$ $(k > 1)$ at a critical metric $g = \bar{g}_0$, and its restriction to the $\mathcal{F}$-conformal variations $S = s\hat{g}$ $(s : M \to \mathbb{R})$ are given by (2.24), where $S = \partial_t \bar{g}_t$, $f = \sigma_k$, $\Phi_f = \frac{1}{4}(n-k+1)(n-k+2)\sigma_{k-2}$, and

$$\Phi_1(S^\sharp) = \frac{1}{4}(\sigma_k - I_{\sigma,k}(g))(\operatorname{Tr} S^\sharp)\,\widehat{\operatorname{id}} - \frac{1}{2}T_{k-1}(A)[S^\sharp, A],$$

$$\Phi_2(S^\sharp) = \frac{1}{4}\sum_{j=0}^{k-2}(-1)^j\left(\operatorname{Tr}(T_{k-j-2}(A)S^\sharp)A^j - \sigma_{k-j-2}\sum_{p=0}^{j} A^p S^\sharp A^{j-p}\right),$$

$$\Phi_3(S^\sharp) = -\frac{1}{2}(\operatorname{Tr} S^\sharp)\,T_{k-1}(A).$$

Proof. Using only Proposition 2.2 and the identity $\operatorname{Tr} T_k(A) = (n-k)\sigma_k$ we obtain $B_f = \frac{1}{2}T_{k-1}(b)$. As in the proof of Theorem 2.1 (see (2.25) with $f = \sigma_k$), we get

$$I_{\sigma,k}'(\bar{g}_t) = \int_M \left(\partial_t \sigma_k + \frac{1}{2}(\sigma_k - I_{\sigma,k}(\bar{g}_t))\operatorname{Tr} S^\sharp\right) d\overline{\operatorname{vol}}_t.$$

By $(2.20)_2$ and Lemma 2.6, we have at $t = 0$

$$\int_M (\partial_t \sigma_k)\,d\overline{\operatorname{vol}} = -\frac{1}{2}\int_M \langle T_{k-1}(A), \nabla_N S^\sharp \rangle d\overline{\operatorname{vol}} = -\frac{1}{2}\int_M \langle \mathcal{V}(T_{k-1}(A)), S^\sharp \rangle d\overline{\operatorname{vol}}.$$

The above (or Theorem 2.1 with $B_f = \frac{1}{2}T_{k-1}(b)$ and $\mathrm{Tr}\, B_f^\sharp = \frac{1}{2}(n-k+1)\sigma_{k-1}$) yield the formula for the gradient $\nabla I_{\sigma,k}(g)$. Concerning the second variation of $I_{\sigma,k}$, as in the proof of Theorem 2.1 (with $f = \sigma_k$) or by (2.28), one has

$$I''_{\sigma,k}(\bar{g}_t)_{|t=0} = \frac{1}{2}\int_M \Big\langle (\partial_t\sigma_k)\,\hat{g} + (\sigma_k - I_{\sigma,k}(g))S - \partial_t\mathcal{V}(T_{k-1}(b)),\ S\Big\rangle d\,\mathrm{vol}. \quad (2.33)$$

We have

$$\partial_t\mathcal{V}(T_{k-1}(b)) = (\partial_t\tau_1)T_{k-1}(b) + \tau_1\partial_t T_{k-1}(b) - \partial_t\nabla_N^t T_{k-1}(b),$$

where $\partial_t T_{k-1}(b)$ and $\partial_t\nabla_N^t T_{k-1}(b)$ are given in Example 2.3(b). Using the identities

$$\mathcal{V}(T_{k-1}(A)) = (\sigma_k - I_{\sigma,k}(g))\widehat{\mathrm{id}} \quad \text{(at a critical metric)},$$

$$\frac{1}{2}\mathrm{Tr}\,([[T_{k-1}(A), S^\sharp], A]\,S^\sharp) = \mathrm{Tr}\,(T_{k-1}(A)[S^\sharp, A]\,S^\sharp),$$

from (2.33), as in the proof of Theorem 2.1, we obtain

$$\begin{aligned}
I''_{\sigma,k}(\bar{g}_t)_{|t=0} = \frac{1}{2}\int_M \Bigg(& \frac{1}{2}(\sigma_k - I_{\sigma,k}(g))(\mathrm{Tr}\,S^\sharp)^2 - \mathrm{Tr}\,(T_{k-1}(A)[S^\sharp, A]S^\sharp) \\
& - \mathrm{Tr}\,(T_{k-1}(A)\nabla_N S^\sharp)\,\mathrm{Tr}\,S^\sharp + \frac{1}{2}\sum_{j=0}^{k-2}(-1)^j\,\mathrm{Tr}\,(T_{k-j-2}(A)\nabla_N S^\sharp)\,\mathrm{Tr}\,(A^j\nabla_N S^\sharp) \\
& + \frac{1}{2}\sum_{j=1}^{k-1}(-1)^j\sigma_{k-j-1}\sum_{p=0}^{j-1}\mathrm{Tr}\,(A^p(\nabla_N^t S^\sharp)A^{j-p-1}(\nabla_N S^\sharp)) \Bigg) d\,\mathrm{vol}.
\end{aligned}$$

The formulae for Φ_i ($i = 1,2,3$) follow from the above. $\square$

Although the Ricci tensor is the notion of intrinsic geometry, the functional

$$E_N(g) = \int_M \mathrm{Ric}(N,N)\,d\,\mathrm{vol}_g, \qquad g \in \mathcal{M}$$

(the total *normal Ricci curvature*) belongs to extrinsic geometry of a foliation $\mathcal{F}$ on (M,g): by known integral formula (1.3) we have $E_N = I_{\sigma,2}$.

Example 2.5. For the function $f = \sigma_2$, we have the following particular case of Corollary 2.3. The gradient of the functional $E_N : \mathcal{M} \to \mathbb{R}$ and its projection via $\pi_* : T\mathcal{M} \to T\mathcal{M}_1$ are given by

$$\bar{\nabla}E_N(g) = \frac{1}{2}(\sigma_2 - E_N(g))\,\hat{g} - \frac{1}{2}\mathcal{V}(T_1(b)).$$

Recall that $T_1(b) = \sigma_1\hat{g} - \hat{b}_1$. The $\mathcal{F}\mathcal{M}_1$- and $\mathcal{F}\mathcal{M}$- components of the above gradient are

$$\nabla^{\mathscr{F}} E_N(g) = \frac{1}{2}\left(\underline{\sigma_2 - E_N(g)} - \frac{n-1}{n}\mathscr{V}(\sigma_1)\right)\hat{g}.$$

The second variation of E_N at a critical metric $g = \bar{g}_0$, and its restriction to the $\mathscr{F}$-conformal variations are given by (2.24), where $S = \partial_t \bar{g}_t$, $f = \sigma_2$, and

$$\Phi_1(S^{\sharp}) = \frac{1}{4}(\underline{\sigma_2 - E_N(g)})(\mathrm{Tr}\, S^{\sharp})\,\widehat{\mathrm{id}} - \frac{1}{2}T_1(A)[S^{\sharp}, A],$$

$$\Phi_2(S^{\sharp}) = \frac{1}{4}(\mathrm{Tr}\, S^{\sharp})\,\widehat{\mathrm{id}} - \frac{1}{4}S^{\sharp}, \qquad \Phi_3(S^{\sharp}) = -\frac{1}{2}(\mathrm{Tr}\, S^{\sharp})T_1(A).$$

Notice that the form $\langle \Phi_2(\nabla_N S^{\sharp}), \nabla_N S^{\sharp}\rangle$ is not definite. By Lemmas 2.7 and 2.8, if the lengths of N-curves are unbounded, there are no stable critical metrics for $E_N(g)$.

Example 2.6. The property "being umbilical" (or being close, in some sense, to such), relates to the measure of "nonumbilicity" for foliations, see [30]. The measure of "nonumbilicity" for foliations is expressed by the functional (2.1) with

$$f = \sum_{i<j}(k_i - k_j)^2 = n\tau_2 - \tau_1^2.$$

Metrics with minimal total "nonumbilicity" (if they exist) are critical for the functional $U_{\mathscr{F}}(g) = \int_M (n\tau_2 - \tau_1^2)\,\mathrm{d\, vol}$. Using notations of Theorem 2.1, one has $B_f = n\hat{b}_1 - \tau_1\hat{g}$. (Notice that $\mathrm{Tr}\, B_f = 0$). In this case,

$$\nabla U_{\mathscr{F}} = \left(\mathscr{V}(\tau_1) + \frac{1}{2}(n\tau_2 - \tau_1^2 - \underline{U_{\mathscr{F}}(g)})\right)\hat{g} - n\mathscr{V}(\hat{b}_1).$$

We also have

$$\Phi_1(S^{\sharp}) = \frac{1}{4}(\mathrm{Tr}\, S^{\sharp})\left(n\tau_2 - \tau_1^2 - \underline{U_{\mathscr{F}}(g)}\right)\widehat{\mathrm{id}} - nA\,[S^{\sharp}, A],$$

$$\Phi_2(S^{\sharp}) = \frac{n}{2}S^{\sharp}, \qquad \Phi_3(S^{\sharp}) = -(\mathrm{Tr}\, S^{\sharp})(nA - \tau_1\,\widehat{\mathrm{id}}).$$

Hence, $\langle \Phi_2(\nabla_N S), \nabla_N S\rangle \geq 0$. By Lemma 2.7(i), if the lengths of N-curves on M are unbounded then $\mu(\Phi_2)$.

2.4 Applications and Examples

2.4.1 *Variational Formulae for Umbilical Foliations*

Let $\mathscr{F}$ be an *umbilical* foliation on (M, g) with the normal curvature $\lambda : M \to \mathbb{R}$. One may show that $\mathscr{F}$-conformal variations $g_t \in \mathscr{M}$ preserve this property (i.e., $\lambda = H = \frac{1}{n}\tau_1$).

Proposition 2.4. *Let $\mathscr{F}$ be an umbilical foliation on (M, g_0). If $g_t \in \mathscr{M}$ $(0 \leq t < \varepsilon)$ is an $\mathscr{F}$-conformal variation of g_0 then $\mathscr{F}$ is umbilical for any g_t.*

Proof. The claim follows from Lemma 2.3 (see also Lemma 2.9 for $S = s\hat{g}$). □

Given function $\psi \in C(\mathbb{R})$, consider the functional $I_\psi : \mathscr{M}_{|\mathscr{U}} \to \mathbb{R}$ on the space $\mathscr{U}$ of all Riemannian metrics with respect to which $\mathscr{F}$ is umbilical,

$$I_\psi(g) = \int_M \psi(\lambda)\, d\mathrm{vol}_g. \tag{2.34}$$

Corollary 2.4. *Let $\mathscr{F}$ be an umbilical foliation on (M, g) with the normal curvature λ, and $\psi \in C^2(\mathbb{R})$. The $\mathscr{F}\mathscr{M}_1$- and $\mathscr{F}\mathscr{M}$- components of the gradient of the functional I_ψ are given by:*

$$\nabla^{\mathscr{F}} I_\psi(g) = \frac{1}{2}\left(\psi(\lambda) \underline{- I_\psi(g)} - \frac{1}{n}\mathscr{V}(\psi'(\lambda))\right)\hat{g}. \tag{2.35}$$

The second variation of I_ψ at a critical metric $g = \bar{g}_0 \in \mathscr{M}_1$, with respect to $\mathscr{F}$-conformal (i.e., of $T\mathscr{F}\mathscr{M}$ or $T\mathscr{F}\mathscr{M}_1$) variations $\bar{g}_t \in \mathscr{M}_1$ with $S = s\hat{g}$ ($s \in C^1(M)$) is

$$I_\psi''(\bar{g}_t)_{|t=0} = \frac{1}{4}\int_M \psi''(\lambda)N(s)^2\, d\mathrm{vol}. \tag{2.36}$$

Proof. Because $\tau_i = n\lambda^i$, we set $f = \psi(\tau_1/n)$ and apply Theorem 2.1. In this case, $B_f = \frac{1}{2n}\psi'\hat{g}$, and, see (2.35),

$$I_\psi'(\bar{g}_t) = \frac{1}{2}\int_M s(\psi \underline{- I_\psi(g)} - \frac{1}{n}\mathscr{V}(\psi'))\, d\overline{\mathrm{vol}}.$$

We prove (2.36) directly. To find I_ψ'' we differentiate the above and get

$$I_\psi''(\bar{g}_t) = \frac{1}{2}\int_M s\left(n\,\partial_t(\psi') - \partial_t(\mathscr{V}(\psi'))\right)\, d\overline{\mathrm{vol}}. \tag{2.37}$$

Using $\partial_t\lambda = -\frac{1}{2}N(s)$, $\partial_t(\psi) = -\frac{1}{2}\psi'N(s)$, and so on, we compute

$$\partial_t(\mathscr{V}(\psi')) = \frac{1}{2}\left((-n\psi' - n\lambda\,\psi'' + N(\psi''))\,sN(s) + \psi''sN(N(s))\right).$$

Hence, the components of the integral (2.37) are

$$\int_M s\,\partial_t(\mathscr{V}(\psi'))\, d\overline{\mathrm{vol}} = \int_M \left(\frac{1}{2}\psi''N(s)^2 - \frac{n}{4}\mathscr{V}(\psi')s^2\right) d\overline{\mathrm{vol}},$$

$$\int_M sn\,\partial_t(\psi')\, d\overline{\mathrm{vol}} = -\frac{n}{4}\int_M \mathscr{V}(\psi')s^2\, d\overline{\mathrm{vol}}.$$

This yields the formula for I_ψ''.   □

Remark 2.6. If $\psi'' > 0$ then the functional $I_\psi : \mathcal{M}_1 \to \mathbb{R}$ restricted to umbilical metrics has at most one critical point. See also Remark 2.5.

2.4.2 The Energy and Bending of the Unit Normal Vector Field

The *energy of a unit vector field N* on (M^{n+1}, g) can be expressed by the formula

$$\mathcal{E}_N(g) = \frac{1}{2}(n+1)\,\mathrm{vol}(M,g) + \int_M \|\nabla N\|^2 \,\mathrm{d\,vol}_g,$$

see, for example, [10]. The last integral,

$$\mathcal{B}_N(g) = \int_M \|\nabla N\|^2 \,\mathrm{d\,vol}_g,$$

up to the constant $c_{n+1} = \frac{1}{n}\mathrm{vol}(S^{n+1}(1))$, is called the *total bending* of N. The problems of minimizing $\mathcal{E}_N(g)$ and $\mathcal{B}_N(g)$ with respect to variations $\bar{g}_t \in \mathcal{M}_1$ are equivalent.

Let $e_0 = N$ and $e_1, \dots, e_n$ be a local orthonormal basis of $(M, \mathcal{F}, g)$. We calculate

$$\|\nabla N\|^2 = \sum_{i=1}^n g(\nabla_{e_i} N, \nabla_{e_i} N) + g(Z, Z) = \tau_2 + \|Z\|^2,$$

where $Z = \nabla_N N$ for short. Thus, we decompose the bending into two parts,

$$\mathcal{B}_N = I_{\tau,2} + \mathcal{B}_N^{\perp}, \quad \text{where} \quad \mathcal{B}_N^{\perp}(g) = \int_M \|Z\|^2 \,\mathrm{d\,vol}.$$

Notice that $\mathcal{B}_N^{\perp} = 0$ for Riemannian foliations, i.e., $Z = 0$.

Lemma 2.10. *The vector field $Z = \nabla_N^t N$ is evolved by $g_t \in \mathcal{M}$ with $S = \partial_t g_t$ as*

$$(i)\ \partial_t Z = -S^{\sharp}(Z), \qquad (ii)\ \partial_t Z = -sZ \quad \text{for} \quad S = s\hat{g}. \tag{2.38}$$

In particular, all variations $g_t \in \mathcal{M}$ preserve Riemannian foliations.

Proof. We use (2.11) to compute for any $X \in T\mathcal{F}$

$$g_t(\partial_t Z, X) = \frac{1}{2}\left(2(\nabla_N^t S)(X, N) - (\nabla_X^t S)(N, N)\right) = -S(\nabla_N^t N, X) = -g_t(S^{\sharp}(Z), X).$$

From this, all of (2.38) follow. If $Z = 0$ at $t = 0$ then by uniqueness of a solution to the linear ODE (2.38)(i) along N-curves, we have $Z = 0$ for all t. $\qquad\square$

By Lemma 2.10, we have $\partial_t(Z^{\flat}) = 0$. Indeed, we calculate (for any vector X)

$$\partial_t(Z^{\flat})(X) = \partial_t(g(Z, X)) = S(Z, X) + g(\partial_t Z, X) = 0.$$

Notice that $\langle Z^{\flat} \odot Z^{\flat}, S \rangle = S(Z, Z)$, in particular, $\langle Z^{\flat} \odot Z^{\flat}, \hat{g} \rangle = \|Z\|^2$.

As the components of 1-form $Z^\flat = g(Z, \cdot)$ are $(Z^\flat)_i = Z^a g_{ia}$, by definition of the tensor product, we have

$$(Z^\flat \odot Z^\flat)_{ij} = Z^a g_{ia} Z^b g_{jb}.$$

From this we obtain

$$\langle Z^\flat \odot Z^\flat, S \rangle = Z^a g_{ia} Z^b g_{jb} S^{ij} = Z^a Z^b S_{ab} = S(Z,Z).$$

By Lemma 2.10, we have $\partial_t (Z^\flat \odot Z^\flat) = 0$. Indeed, we find (for any vectors X,Y)

$$\begin{aligned}
\partial_t (Z^\flat \odot Z^\flat)(X,Y) &= \partial_t (g(Z,X)g(Z,Y)) \\
&= S(Z,X)g(Z,Y) + g(Z,X)S(Z,Y) + g(\partial_t Z,X)g(Z,Y) \\
&\quad + g(Z,X)g(\partial_t Z,Y) = 0.
\end{aligned}$$

Theorem 2.2. *The gradient of the bending functional $\mathcal{B}_N : \mathcal{M} \to \mathbb{R}$ (and its projection via $\pi_* : T\mathcal{M} \to T\mathcal{M}_1$) is given by*

$$\nabla \mathcal{B}_N(g) = \frac{1}{2} \left(\|Z\|_g^2 + \tau_2 - \mathcal{B}_N(g) \right) \hat{g} - Z^\flat \odot Z^\flat - \mathcal{V}(\hat{b}_1),$$

where $Z = \nabla_N^t N$. The $\mathcal{F}\mathcal{M}_1$- (and $\mathcal{F}\mathcal{M}$-) component of the gradient is

$$\nabla^{\mathcal{F}} \mathcal{B}_N(g) = \left(\frac{1}{2} \left(\|Z\|_g^2 + \tau_2 - \mathcal{B}_N(g) \right) - \frac{1}{n} \|Z\|_g^2 - \frac{1}{n} \mathcal{V}(\tau_1) \right) \hat{g}.$$

The second variation of $\mathcal{B}_N$ at a critical metric $g = \bar{g}_0$, where $S = \partial_t \bar{g}_t$, and its restriction to the $\mathcal{F}$-conformal variations (i.e., $S = s\hat{g}$, $s : M \to \mathbb{R}$) are, respectively,

$$\mathcal{B}_N''(\bar{g}_t)_{|t=0} = \int_M \left(\langle \Phi_1(S), S \rangle + \langle \Phi_2(\nabla_N S), \nabla_N S \rangle + \langle \Phi_3(S), \nabla_N S \rangle \right) d\mathrm{vol},$$

$$\bar{\mathcal{B}}_N''(\bar{g}_t)_{|t=0} = \frac{n}{2} \int_M \left(\left(\frac{n}{2} - 1 \right) \|Z\|_g^2 s^2 + N(s)^2 \right) d\mathrm{vol},$$

where

$$\Phi_1(S^\sharp) = -\frac{1}{4} \left(\|Z\|_g^2 + \tau_2 - \mathcal{B}_N(g) \right) (\mathrm{Tr}\, S^\sharp) \widehat{\mathrm{id}} - A[S^\sharp, A] - S^2(Z,Z) \widehat{\mathrm{id}},$$

$$\Phi_2(S^\sharp) = \frac{1}{2} S^\sharp, \qquad \Phi_3(S^\sharp) = \mathrm{Tr}\,(A S^\sharp) \widehat{\mathrm{id}}.$$

Proof. First, using (2.11) and case (i) in Lemma 2.10, we compute

$$\partial_t \|Z\|_{g_t}^2 = (\partial_t g_t)(Z,Z) + 2 g_t(\partial_t Z,Z) = S(Z,Z) - 2 S(Z,Z) = -S(Z,Z).$$

For $f = \|Z\|_{g_t}^2 + \tau_2$ we have

$$\partial_t f = -S(Z,Z) - \mathrm{Tr}\,(A\nabla_N^t S^\sharp).$$

Hence

$$\int_M (\partial_t f)\,\mathrm{d\,vol} = -\int_M \left(S(Z,Z) + \langle \mathscr{V}(\hat{b}_1),\, S\rangle\right)\mathrm{d\,vol}.$$

Then, similarly to (2.25) and (2.27), we obtain

$$\mathscr{B}_N'(\bar{g}_t) = \int_M \left\langle \frac{1}{2}\left(\|Z\|_{g_t}^2 + \tau_2 - \mathscr{B}_N(\bar{g}_t)\right)\hat{g} - Z^\flat \odot Z^\flat - \mathscr{V}(\hat{b}_1),\, S\right\rangle \mathrm{d\,vol}. \quad (2.39)$$

In order to find the second variations, using Lemma 2.4, as for (2.28) we compute

$$\mathscr{B}_N''(\bar{g}_t)|_{t=0} = \int_M \left(\frac{1}{2}(\|Z\|_g^2 + \tau_2 - \mathscr{B}_N(g))\langle S,S\rangle + \mathrm{Tr}\,(A\nabla_N^t S^\sharp)\right)(\mathrm{Tr}\,S^\sharp)$$

$$-\frac{1}{2}\left(S(Z,Z) - \langle \partial_t \mathscr{V}(\hat{b}_1),\, S\rangle\right)\mathrm{d\,vol}.$$

We have $\partial_t \mathscr{V}(\hat{b}_1) = (\partial_t \tau_1)\hat{b}_1 + \tau_1\partial_t\hat{b}_1 - \partial_t\nabla_N^t\hat{b}_1$, where by Example 2.3(a),

$$\int_M \langle \tau_1 \partial_t \hat{b}_1, S\rangle \mathrm{d\,vol} = \int_M \tau_1\left(\mathrm{Tr}\,(S^\sharp A S^\sharp) - \frac{1}{4}N\left(\mathrm{Tr}\,(S^{\sharp 2})\right)\right)\mathrm{d\,vol},$$

$$\int_M \langle \partial_t(\nabla_N \hat{b}_1),\, S\rangle \mathrm{d\,vol} = \int_M \left(\mathrm{Tr}\,\left(S^\sharp(\nabla_N^t A)S^\sharp\right) - \mathrm{Tr}\,(A[S^\sharp, A]S^\sharp)\right.$$

$$\left. -\frac{1}{4}\tau_1 N(\mathrm{Tr}\,(S^{\sharp 2})) + \frac{1}{2}\mathrm{Tr}\,((\nabla_N^t S^\sharp)^2)\right)\mathrm{d\,vol}.$$

Hence

$$\int_M \langle \partial_t \mathscr{V}(\hat{b}_1), S\rangle \mathrm{d\,vol} = \int_M \left(-\frac{1}{2}N(\mathrm{Tr}\,S^\sharp)\,\mathrm{Tr}\,(AS^\sharp) + \mathrm{Tr}\,(S^\sharp \mathscr{V}(A)S^\sharp)\right.$$

$$\left. + \mathrm{Tr}\,(A[S^\sharp, A]S^\sharp) - \frac{1}{2}\mathrm{Tr}\,((\nabla_N^t S^\sharp)^2)\right)\mathrm{d\,vol}.$$

Finally, we obtain (see also Corollary 2.2 for $k = 2$)

$$\mathscr{B}_N''(\bar{g}_t)|_{t=0} = \int_M \left(-\frac{1}{4}\left(\|Z\|_g^2 + \tau_2 - \mathscr{B}_N(g)\right)(\mathrm{Tr}\,S^\sharp)^2 - \mathrm{Tr}\,(A[S^\sharp, A]S^\sharp)\right.$$

$$\left. -S^2(Z,Z) + \mathrm{Tr}\,(AS^\sharp)N(\mathrm{Tr}\,S^\sharp) + \frac{1}{2}\mathrm{Tr}\,((\nabla_N S^\sharp)^2)\right)\mathrm{d\,vol}.$$

Formulae for $\mathscr{F}$-conformal case follow directly from above.   □

Chapter 3
Extrinsic Geometric Flows

Abstract In the chapter we study the metrics g_t satisfying the *Extrinsic Geometric Flow* equation (see Sect. 3.2). Sections 3.4 and 3.5 collect results about existence and uniqueness of solutions (Theorems 3.1 and 3.2) and their proofs. The key role in proofs play hyperbolic PDEs and the generalized companion matrix studied in Sect. 3.3. In Sect. 3.6, we estimate the maximal existence time. In Sect. 3.7 we use the first derivative of functionals (when they are monotonous) to show convergence of metrics in a weak sense (Theorem 3.3). In Sect. 3.8 we study soliton solutions of the geometric flow equation (Theorems 3.4 and 3.5), and characterize them in the cases of umbilical foliations and foliations on surfaces (Theorems 3.6–3.8). Section 3.9 is devoted to applications and examples, including the geometric flow produced by the extrinsic Ricci curvature tensor (Theorem 3.9).

3.1 Introduction

Geometric Flow (GF) is an evolution equation associated to a functional on a manifold which has a geometric interpretation, usually related to some extrinsic or intrinsic curvature. They all correspond to dynamical systems in the infinite dimensional space of all possible metrics on a given manifold,

$$\partial_t g_t = h(g_t).$$

These evolutions try to move g_0 toward a metric that is more natural for its underlined topology (e.g., with constant curvature). GF equations are quite difficult to solve in all generality, because of their nonlinearity. Although the short time existence of solutions is guaranteed by the parabolic or hyperbolic nature of the equations, their (long time) convergence to canonical metrics is analyzed under various conditions (e.g., in connection to the problem of formation of singularities).

The theory of intrinsic (driven by curvature in various forms) and extrinsic (driven by extrinsic curvature of submanifolds) GFs is a modern subject of common critical interest in mathematics and physics. The most popular GF in

V. Rovenski and P. Walczak, *Topics in Extrinsic Geometry of Codimension-One Foliations*, SpringerBriefs in Mathematics, DOI 10.1007/978-1-4419-9908-5_3,
© Vladimir Rovenski and Paweł Walczak 2011

mathematics are the *Ricci flow* and the *Mean Curvature flow*. Other examples are Gaussian curvature flow, Yamabe flow, Renormalization group flow, etc. The physical applications of GF include problems in quantum field theory as well as problems in fluid mechanics, general relativity, and string theory.

Some of the most striking recent results in differential geometry and topology are related to the *Ricci flow*, that is the deformation g_t of a given Riemannian metric g_0 on a manifold M subject to the equation

$$\partial_t g_t = -2\operatorname{Ric}_t,$$

where Ric_t is the Ricci curvature tensor on the Riemannian manifold (M, g_t). This evolution equation has been introduced by Hamilton [23] (following earlier work of Eells and Sampson [19] on the harmonic map heat flow, see also [5]) who proposed its use in solving famous Poincaré and Geometrization Conjectures. Using the Ricci flow, Hamilton proved that every compact three-dimensional manifold with positive Ricci curvature is diffeomorphic to a spherical space form. Hamilton's program was completed by Perelman who proved these conjectures in the series of preprints [35–37]. As some authors claim "Perelman's proofs are concise and, at times, sketchy", several authors made efforts to provide details that are missing in Perelman's preprints, or to present "a proof more attractive to topologists", see, for example, [7, 25]. Also, there exist several successful attempts to modify/simplify proofs of some elements constituting the proof of the Conjectures.

Discussing these proofs is not our goal here (equations describing our extrinsic geometric flows (EGFs) are quite different from those related to the Ricci flow), these days interested readers can find enormous amount of articles (especially at arXiv) and books on Ricci flow, [2,7,8,13,14,53], and so on.

A famous (Rauch, Klingenberg, and Berger) theorem states that a complete simply connected n-dimensional Riemannian manifold, for which the sectional curvatures are strictly between 1 and 4, is homeomorphic to n-sphere. It has been a longstanding open conjecture as to whether or not the homeomorphism conclusion could be strengthened to a diffeomorphism. Only recently, Brendle-Schoen proved this hypothesis, see [2, 8]. Their proof is based on analysis of Ricci flow including: Perelman's monotonicity formulae, the blow-up analysis of singularities, and development of recent convergence theory for the Ricci flow.

On the other hand, for at least 30 years there has been continuous interest in the study of *Mean Curvature Flow*, i.e., the variation of immersions $F_0 : \bar{M} \to M$ of manifolds $\bar{M}$ into Riemannian manifolds (M, g) subject to the *mean curvature vector field*, i.e., to the equation

$$\partial_t F_t = H_t,$$

where H_t is the mean curvature vector of $F_t(\bar{M})$, see [18,21]. Also, the second author [55] considered the foliated version of the Mean Curvature Flow: foliations which are invariant under the flow of the mean curvature vector of their leaves.

The main goal of Chap. 3 is the study of *Extrinsic Geometric Flows* on foliations. They are defined as deformations of Riemannian metrics on a manifold M equipped with a codimension-one foliation subject to conditions expressed in terms of the second fundamental form of the leaves and its scalar invariants.

The main results of the chapter are the local existence/uniqueness theorems, estimations of the existence time of solutions, the convergence in a weak sense to minimal, and totally geodesic foliations. Other results concern the geometry of extrinsic Ricci and Newton transformation flows, the introducing of geometric solitons and their classification among umbilical foliations and metrics on foliated surfaces.

3.2 The Systems of PDEs Related to EGFs

We study two types of evolution of Riemannian structures, depending on functions f_j $(0 \le j < n)$ (at least one of them is not identically zero):

(a) $f_j \in C^2(M \times \mathbb{R})$,
(b) $f_j = \tilde{f}_j(\vec{\tau}, \cdot)$, where $\tilde{f}_j \in C^2(\mathbb{R}^{n+1})$.

Sometimes we will assume that $f_j = \tilde{f}_j(\vec{\tau})$ with $\tilde{f}_j \in C^2(\mathbb{R}^n)$.

Definition 3.1. Given functions f_j of type (a) or (b), a family g_t, $t \in [0, \varepsilon)$, of Riemannian structures on $(M, \mathscr{F})$ will be called an *Extrinsic Geometric Flow* (EGF) if

$$\partial_t g_t = h_t, \quad \text{where } h_t = h(b_t) = \sum_{j=0}^{n-1} f_j \hat{b}_j^t. \tag{3.1}$$

Here, $\hat{b}_j^t$ are symmetric $(0,2)$-tensor fields on M g_t-dual to $(A_t)^j$ (See Sect. 1.2.2).

The choice of the right-hand side in (3.1) for $h(b)$ seems to be natural, the powers $\hat{b}_j$ are the only $(0,2)$-tensors which can be obtained algebraically from the second fundamental form b, while $\tau_1, \ldots, \tau_n$ (or, equivalently, $\sigma_1, \ldots, \sigma_n$) generate all the scalar invariants of extrinsic geometry.

Powers $\hat{b}_j$ with $j > 1$ in (3.1) are meaningful; for example, the EGFs produced by:

- The extrinsic Ricci curvature tensor $\mathrm{Ric}^{\mathrm{ex}}(b)$, see (3.87), depends on $\hat{b}_1$ and $\hat{b}_2$.
- The Newton transformation $T_i(b) = T_i(A)^\flat$, see (1.6), depends on all $\hat{b}_j$ $(j \le i)$.

In other words, the EGF is the evolution equation which deforms Riemannian metrics by evolving them along $\mathscr{F}$ in the direction of the tensor $h(b_t)$. Indeed, any EGF preserves N to be unit and perpendicular to $\mathscr{F}$, therefore, the $\mathscr{F}$-component of the vector does not depend on t.

One can interpret EGF as integral curves of a vector field $g \mapsto h(b), h(b)$ being the right-hand side of (3.1), on the space $C^k(M, S_2^+(M))$ of Riemannian C^k-structures on M. Here, certainly, $S_2^+(M)$ is the bundle of positive definite symmetric $(0,2)$-tensors on M. This vector field may or may not depend on time.

Although EGF of type (b) consists of first-order nonlinear PDEs, the corresponding power sums τ_i $(i > 0)$ satisfy an infinite quasilinear system.

Lemma 3.1. *Power sums* $\{\tau_i\}_{i\in\mathbb{N}}$ *of the EGF (3.1) of type (b) satisfy the system*

$$\partial_t \tau_i = -\frac{i}{2}\left\{\tau_{i-1}N(f_0) + \sum_{j=1}^{n-1}\left(\frac{jf_j}{i+j-1}N(\tau_{i+j-1}) + \tau_{i+j-1}N(f_j)\right)\right\}, \quad (3.2)$$

where $N(f_j) = \sum_{s=1}^{n} f_{j,\tau_s}N(\tau_s)$*. Moreover, the evolution of* σ_k $(k \leq n)$ *is given by*

$$2\partial_t \sigma_k = -\sum_{i=0}^{k-1}(-1)^i \sigma_{k-i-1}\left\{N(f_0)\,\tau_i + \sum_{j=1}^{n-1}\left(N(f_j)\,\tau_{i+j} + \frac{j}{i+j}f_j N(\tau_{i+j})\right)\right\}. \tag{3.3}$$

Proof. For the EGF (3.1), the equality (2.20)$_1$ with $S = h(b)$ reduces to the system

$$\partial_t \tau_i = -\frac{i}{2}\,\mathrm{Tr}\left(A^{i-1}\nabla_N^t\left(\sum_{j=0}^{n-1}f_j A^j\right)\right), \quad i > 0.$$

The desired (3.2) follows from the above, using the identity

$$N(\tau_{i+j}) = \mathrm{Tr}(\nabla_N^t A^{i+j}) = \frac{i+j}{j}\,\mathrm{Tr}(A^i \nabla_N^t A^j). \tag{3.4}$$

We also prove (3.2) directly in what follows, see Remark 3.3 in Sect. 3.5.1.
From (2.20)$_2$ with $S = h(b)$, we find

$$2\partial_t \sigma_k = -\mathrm{Tr}\left(T_{k-1}(A)\nabla_N^t\left(\sum_{j=0}^{n-1}f_j A^j\right)\right).$$

From the above, using the identity (3.4), we obtain (3.3). □

Example 3.1. We define the *k*th *Newton transformation flow* as:

$$\partial_t g_t = T_k(b_t). \tag{3.5}$$

In other words, we assume $f_j = (-1)^j \sigma_{k-j}$ in (3.1).
For the flow (3.5), the equality (2.20)$_1$ with $S = T_k(b)$ reduces to the system

$$\partial_t \tau_i = -\frac{i}{2}\,\mathrm{Tr}(A^{i-1}\nabla_N^t T_k(A)), \quad i > 0.$$

Using the identity (3.4), we obtain the system of PDEs of type (3.2),

$$\partial_t \tau_i = -\frac{i}{2}\,\mathrm{Tr}\left(A^{i-1}\nabla_N^t\left(\sum_{j=0}^{k}(-1)^j\sigma_{k-j}A^j\right)\right)$$

$$= -\frac{i}{2}\sum_{j=0}^{k}(-1)^j\left(N(\sigma_{k-j})\tau_{i+j-1} + \sigma_{k-j}\,\mathrm{Tr}(A^{i-1}\nabla_N(A^j))\right)$$

$$= -\frac{i}{2}\left\{N(\sigma_k)\tau_{i-1} + \sum_{j=1}^{k}(-1)^j\left(N(\sigma_{k-j})\tau_{i+j-1} + \frac{j}{i+j-1}\sigma_{k-j}N(\tau_{i+j-1})\right)\right\}.$$

From $(2.20)_2$ with $S = T_k(b)$ we also obtain the system of PDEs for σ's

$$\partial_t \sigma_i = -\frac{1}{2}\mathrm{Tr}\,(T_{i-1}(A)\nabla_N^t T_k(A)), \quad 0 < i \leq n. \tag{3.6}$$

In particular, for $i = 1$, we have

$$\partial_t \sigma_1 = -\frac{1}{2}\mathrm{Tr}\,(\nabla_N^t T_k(A)) = -\frac{1}{2}(n-k)N(\sigma_k).$$

For $k = 1$, (3.6) reduces to the system (for $0 < i \leq n$)

$$\partial_t \sigma_i = -\frac{1}{2}\mathrm{Tr}\,(T_{i-1}(A)\nabla_N^t T_1(A)) = -\frac{1}{2}(n-i+1)\sigma_{i-1}N(\sigma_1) + \frac{1}{2}N(\sigma_i). \tag{3.7}$$

For $i = 1$ we have the linear PDE

$$\partial_t \sigma_1 = -\frac{1}{2}(n-1)N(\sigma_1)$$

representing the "unidirectional wave motion" $\sigma_1^t(s) = \sigma_1^0(s - t(n-1)/2)$ on any N-curve $\gamma : s \mapsto \gamma(s)$. Hence, the (triangular) system (3.7) of PDEs for σ's is solvable.

EGFs preserve the following properties of foliations to be:

- *umbilical* ($A = \lambda\,\widehat{\mathrm{id}}$), see Proposition 3.4.
- *totally geodesic* ($A = 0$), see (2.19).
- *Riemannian* ($\nabla_N N = 0$), see Lemma 2.10.

For a particular choice of functions f_j, EGF might preserve a certain extrinsic geometric property (P) of foliations. For example, by (3.2), EGFs preserve *minimal* ($\tau_1 = 0$) foliations, when the generating functions satisfy the conditions

$$f_j(0, \tau_2, \dots, \tau_n) \equiv 0 \quad (j \neq 1).$$

Let g_t be a solution of (3.1). From (3.2) with $i = 1$ we have

$$2\partial_t \tau_1 = -nN(f_0(\overrightarrow{\tau})) - \sum_{j=1}^{n-1}\left(f_j(\overrightarrow{\tau})N(\tau_j) + \tau_j N(f_j(\overrightarrow{\tau}))\right).$$

One may show that if $\mathscr{F}$ is g_0-minimal then $\partial_t \tau_1 = 0$, i.e., $\mathscr{F}$ is g_t-minimal for all t.

We propose the EGF (3.1) as a tool for studying the following question:
Under what conditions on $(M, \mathscr{F}, g_0)$ the EGF metrics g_t converge to one for which $\mathscr{F}$ enjoys a given extrinsic geometric property (P), e.g., is umbilical, totally geodesic, Riemannian, minimal, etc.?

3.3 Auxiliary Results

3.3.1 Diffeomorphism Invariance of EGFs

It is well known that the Riemannian curvature tensor is invariant under local isometries, and the Ricci curvature tensor Ric is preserved by diffeomorphisms $\phi \in \mathrm{Diff}(M)$ in the sense that $\phi^*(\mathrm{Ric}(g)) = \mathrm{Ric}(\phi^*g)$. Hence, Ric is an *intrinsic geometric tensor*. In the same way, the second fundamental form b of a foliation is invariant under diffeomorphisms preserving the foliation; thus, b is an *extrinsic geometric tensor* for the foliation. More precisely, we have the following

Proposition 3.1. *Let $(M_i, \mathscr{F}_i, g_i)$ $(i = 1, 2)$ be Riemannian manifolds with foliations, and $\phi : M_1 \to M_2$ a diffeomorphism such that $\mathscr{F}_1 = \phi^{-1}(\mathscr{F}_2)$ and $g_1 = (\sigma \cdot \phi^* \hat{g}_2) \oplus \phi^*(g_2^{\perp})$ (for some positive $\sigma \in \mathbb{R}$). Then the Weingarten operators A_i, the second fundamental forms $b^{(i)}$, and $\overrightarrow{\tau}^{(i)}$ (the set of τ-s for A_i) satisfy*

$$b^{(1)} = \sigma \cdot \phi^* b^{(2)}, \qquad A_1 = \phi_*^{-1} A_2 \phi_*, \qquad \overrightarrow{\tau}^{(1)} = \overrightarrow{\tau}^{(2)} \circ \phi. \tag{3.8}$$

For any tensor $h(b)$ of the form $h(b) = \sum_{j=0}^{n-1} f_j(\overrightarrow{\tau}) \hat{b}_j$, see (3.1), we have

$$h(b^{(1)}) = \sigma \cdot \phi^*(h(b^{(2)})). \tag{3.9}$$

Proof. This follows from the following computation (see also Lemma 2.3).

Notice that for any vector $X \in T\mathscr{F}_1$, the vector $\phi_* X = d\phi(X)$ is tangent to $\mathscr{F}_2$. From the definition of the metric g_1, we have $g_1(X, Y) = \sigma \cdot g_2(\phi_* X, \phi_* Y)$ for $X, Y \in T\mathscr{F}_1$. Let N_2 be a unit normal of $\mathscr{F}_2$. Then $N_1 = \phi_*^{-1} N_2$ is a unit normal of $\mathscr{F}_1$. In fact,

$$g_1(X, N_1) = g_2(\phi_* X, N_2) = 0 \qquad (X \in T\mathscr{F}_1),$$

$$g_1(N_1, N_1) = \phi^*(g_2)(\phi_*^{-1} N_2, \phi_*^{-1} N_2) = g_2(\phi_* \phi_*^{-1} N_2, \phi_* \phi_*^{-1} N_2) = g_2(N_2, N_2).$$

For the second fundamental form $b^{(2)} : T\mathscr{F}_2 \times T\mathscr{F}_2 \to \mathbb{R}$ of $\mathscr{F}_2$, we obtain a symmetric $(0,2)$-tensor field $\phi^* b^{(2)} : T\mathscr{F}_1 \times T\mathscr{F}_1 \to \mathbb{R}$. Denote by $\nabla^{(i)}$ the Levi-Civita connection of (M_i, g_i). For arbitrary vector fields $X, Y \subset T\mathscr{F}_1$ we have

$$(\phi^* b^{(2)})(X, Y) = b^{(2)}(\phi_* X, \phi_* Y) = -g_2(\nabla^{(2)}_{\phi_* X} N_2, \phi_* Y) = -g_2(\nabla^{(2)}_{\phi_* X} \phi_* N_1, \phi_* Y)$$

$$= -g_2(\phi_* \nabla^{(1)}_X N_1, \phi_* Y) = -\phi^*(g_2)(\nabla^{(1)}_X N_1, Y) = \sigma^{-1} \cdot b^{(1)}(X, Y).$$

Hence, the second fundamental form of a foliation is an *extrinsic geometric tensor*. Consequently, we have

$$\phi_*(A_1 X) = A_2(\phi_* X) \quad (X \in T\mathscr{F}_1) \quad \Rightarrow \quad A_1 = \phi_* A_2 \phi_*^{-1}.$$

Taking the trace, we get the identity for τ's. For $h(b^{(i)})$ we have

$$\sum_{j=0}^{n-1} f_j(\vec{\tau}^{(1)}) \hat{b}_j^{(1)}(X,Y) = \sigma \sum_{j=0}^{n-1} f_j(\vec{\tau}^{(2)} \circ \phi) \hat{b}_j^{(2)}(\phi_* X, \phi_* Y).$$

From the above, (3.9) follows. □

So if $\phi : (M, \mathscr{F}) \to (\bar{M}, \bar{\mathscr{F}})$ is a diffeomorphism such that $g_t = \phi^* \bar{g}_t$, $\mathscr{F} = \phi^{-1}(\bar{\mathscr{F}})$, and $\bar{g}_t$ is a solution to EGF on $(\bar{M}, \bar{\mathscr{F}})$, by Proposition 3.1 with $\sigma = 1$, we obtain

$$\partial_t g_t' = \phi^*(\partial_t \bar{g}_t) = \phi^* h(\bar{b}_t) = h(b_t).$$

Hence, g_t is also a solution to EGF on $(M, \mathscr{F})$. Therefore, EGF with $h(b) = \sum_{j=0}^{n-1} f_j(\vec{\tau}) \hat{b}_j$ is invariant under the group $\mathscr{D}(\mathscr{F}, N)$ of diffeomorphisms preserving both, the foliation $\mathscr{F}$ and the unit normal N.

We can also translate EGFs along the time coordinate; If g_t satisfies to the EGF equation (3.1), then so does g_{t-t_0} for any t_0. Furthermore, EGF has a *scale invariance* along $\mathscr{F}$; If g_t is a solution to EGF, and $\mu > 0$ is real, then $\bar{g}_t := (\mu \hat{g}_{t/\mu}) \oplus g_{t/\mu}^{\perp}$ is also a solution.

3.3.2 Quasi-Linear PDEs

Now, we recall some facts about hyperbolic systems of quasi-linear PDEs.

Let $A = (a_{ij}(x,t,u))$ be an $n \times n$ matrix, $b = (b_i(x,t,u))$ – an n-vector. A first-order *quasilinear* system of PDEs, n equations in n unknown functions $u = (u_1, \ldots, u_n)$ and two variables $x, t \in \mathbb{R}$, has the form:

$$\partial_t u + A(x,t,u) \partial_x u = b(x,t,u). \tag{3.10}$$

When the coefficient matrix A and the vector b are functions of x and t only, the system is just *linear*; if b alone depends also on u, the system is said to be *semilinear*. The *initial value problem* for (3.10) with given smooth data u_0, A and b consists in finding smooth function $u(x,t)$ satisfying (3.10) and $u(x,0) = u_0(x)$.

Definition 3.2. The system (3.10) is *hyperbolic* in the t-direction at (x,t,u) (in an appropriate domain of the arguments of A and b) if the right eigenvectors of A are real and span $\mathbb{R}^n$. For a solution $u(x,t)$ to (3.10), the corresponding eigenvalues $\lambda_i(x,t,u)$ are called the *characteristic speeds*. The system is *strictly hyperbolic* if the functions $\lambda_i(x,t,u)$ are distinct. For the hyperbolic system (3.10), the vector field $\partial_t + \lambda_i(x,t,u)\partial_x$ is called the *i-characteristic field*, and its integral curves are called *i-characteristics*, see Fig. 3.1.

Remark that the *hyperbolicity of A* is equivalent to any of the properties: "A has real eigenvalues $\lambda_1 \leq \ldots \leq \lambda_n$ and simple elementary divisors (i.e., A has no Jordan cells of order greater than one)" and "A is diagonal in some affine basis". Hence, the

Fig. 3.1 Method of
characteristics

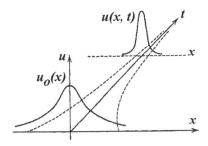

hyperbolic matrix A can be represented as $A = RDR^{-1}$, where R is a nonsingular $n \times n$ matrix and D is a diagonal matrix. The columns r_i of R are the right eigenvectors of A, whereas the rows of L^{-1} are left eigenvectors of A.

A hyperbolic system reduces to the ODEs for its characteristic fields. Indeed, multiplying (3.10) by r_i^T and using $du/dt = \partial_t u + \lambda_i \partial_x u$, we obtain the ODE

$$r_i^T\, du/dt = r_i^T\, b \ \text{ along the characteristic } \ dx/dt = \lambda_i(x,t,u).$$

Theorem A [24]. *Let the quasi-linear system (3.10) be such that*

1) *It is hyperbolic in the t-direction in $\Omega = \{|x| \le a,\ 0 \le t \le s,\ \|u\|_\infty \le r\}$ for some $s, r > 0$.*
2) *the matrix A and the vector b are C^1-regular in Ω.*

If the initial condition

$$u(0,x) = u_0(x), \quad x \in [-a,a] \tag{3.11}$$

has C^1-regular u_0 in $[-a,a]$ and $\|u_0\|_\infty < r$ then (3.10)–(3.11) admit a unique C^1-regular solution $u(x,t)$ in the domain $\bar{\Omega} = \{(x,t):\ |x| + Kt \le a,\ 0 \le t \le \varepsilon\}$, with

$$K = \max\{|\lambda_i(x,t,u)| :\ (x,t,u) \in \Omega,\ 1 \le i \le n\}.$$

Example 3.2. For any function $\psi \in C^1(\mathbb{R})$, we can multiply the equation

$$\partial_t u + \psi(u)\, \partial_x u = 0, \tag{3.12}$$

by $\psi'(u)$, and obtain $\partial_t \psi + \psi \cdot \partial_x \psi = 0$ (the *inviscid Burgers' equation*). Thus, the behavior of the solutions to (3.12) (for t before the first singular value) is not expected to be much different from that of Burgers' equation.

3.3.3 Generalized Companion Matrices

Let $P_n = \lambda^n - p_1 \lambda^{n-1} - \ldots - p_{n-1}\lambda - p_n$ be a polynomial over $\mathbb{R}$ and $\lambda_1 \le \lambda_2 \le \ldots \le \lambda_n$ be the roots of P_n for $n > 0$. Hence, $p_i = (-1)^{i-1}\sigma_i$, where σ_i are elementary symmetric functions of the roots λ_i.

Definition 3.3. Let $c_1 = 1$ and $c_i \neq 0$ $(i > 1)$ be arbitrary numbers. The *generalized companion matrices* of P_n are defined by:

$$
\hat{C} = \begin{pmatrix}
0 & \frac{c_{n-1}}{c_n} & 0 & \cdots & 0 \\
0 & 0 & \frac{c_{n-2}}{c_{n-1}} & \cdots & 0 \\
\cdots & \cdots & \cdots & \cdots & \cdots \\
0 & 0 & \cdots & 0 & \frac{c_1}{c_2} \\
c_n p_n & c_{n-1} p_{n-1} & \cdots & c_2 p_2 & c_1 p_1
\end{pmatrix}
\quad \text{and} \quad
\check{C} = \begin{pmatrix}
c_1 p_1 & c_2 p_2 & \cdots & c_{n-1} p_{n-1} & c_n p_n \\
\frac{c_1}{c_2} & 0 & \cdots & 0 & 0 \\
0 & \frac{c_2}{c_3} & 0 & \cdots & 0 \\
\cdots & \cdots & \cdots & \cdots & \cdots \\
0 & \cdots & 0 & \frac{c_{n-1}}{c_n} & 0
\end{pmatrix}.
$$

Notice that $\hat{C}$ acts on $\mathbb{R}^n$ as $x \to \hat{C}x$, where $x = (x_1, \ldots, x_n)$. Inverting the order of indices, i.e., taking $(x_n, \ldots, x_1)$, one may describe this action by $\check{C}$. If all c_i's are equal to 1, the matrix $\hat{C}$ reduces to the standard *companion matrix* C of P_n. Explicit formulae (polynomials) for entries in powers of C and some applications to the theory of the symmetric functions are given in [17]. The reader can verify that the inverse of the matrix $\hat{C}$ (when $p_n \neq 0$) is $\check{C}$ for $\tilde{P}_n = -\lambda^n p_n^{-1} P_n(\lambda^{-1})$ and $\tilde{c}_i = \frac{c_{n-i+1}}{c_n}$:

$$
\hat{C}^{-1} = \begin{pmatrix}
-\frac{p_{n-1}}{p_n} & -\frac{c_{n-1}}{c_n}\frac{p_{n-2}}{p_n} & \cdots & -\frac{c_2}{c_n}\frac{p_1}{p_n} & \frac{c_1}{c_n}\frac{1}{p_n} \\
\frac{c_n}{c_{n-1}} & 0 & \cdots & 0 & 0 \\
0 & \frac{c_{n-1}}{c_{n-2}} & 0 & \cdots & 0 \\
\cdots & \cdots & \cdots & \cdots & \cdots \\
0 & \cdots & 0 & \frac{c_2}{c_1} & 0
\end{pmatrix}.
$$

The following matrix (the matrix $\hat{C}$ with $c_i = \frac{n}{n-i+1}$) plays a key role in this chapter:

$$
B_{n,1} = \begin{pmatrix}
0 & \frac{1}{2} & 0 & \cdots & 0 \\
0 & 0 & \frac{2}{3} & \cdots & 0 \\
\cdots & \cdots & \cdots & \cdots & \cdots \\
0 & 0 & 0 & \cdots & \frac{n-1}{n} \\
(-1)^{n-1}\frac{n}{1}\sigma_n & (-1)^{n-2}\frac{n}{2}\sigma_{n-1} & \cdots & -\frac{n}{n-1}\sigma_2 & \sigma_1
\end{pmatrix}.
\tag{3.13}
$$

Lemma 3.2. *Generalized companion matrices have the following properties:*

a) *The characteristic polynomial of $\hat{C}$ (or $\check{C}$) is P_n.*

b) $v_j = (1, \frac{c_n}{c_{n-1}}\lambda_j, \frac{c_n}{c_{n-2}}\lambda_j^2, \ldots, c_n\lambda_j^{n-1})$ *is the eigenvector of $\hat{C}$ for the eigenvalue λ_j,*
   respectively, $w_j = (c_n\lambda_j^{n-1}, \ldots, \frac{c_n}{c_{n-2}}\lambda_j^2, \frac{c_n}{c_{n-1}}\lambda_j, 1)$ is the eigenvector of $\check{C}$ for λ_j.

c) $\hat{C}V = VD$, *where $V = \{\frac{c_n}{c_{n-i+1}}\lambda_j^{i-1}\}_{1 \leq i,j \leq n}$ is the Vandermonde type matrix, and*
   $D = \text{diag}(\lambda_1, \ldots, \lambda_n)$ *a diagonal matrix. (If all λ_i's are distinct then obviously*
   $V^{-1}\hat{C}V = D$).

Proof. (a) We show by induction on n that $\det |\lambda \, \text{id}_n - \hat{C}| = P_n$, hence the eigenvalues λ_i of $\hat{C}$ are the roots of P_n. Expanding by co-factors down the first column, we obtain

$$\det |\lambda \,\mathrm{id}_n - \hat{C}| = \lambda \, P_{n-1} - (-1)^{n-1} c_n \, p_n \prod_{i=1}^{n-1} \left(-c_i/c_{i+1} \right),$$

where $P_{n-1} = \lambda^{n-1} - p_1 \lambda^{n-2} - \ldots - p_{n-2}\lambda - p_{n-1}$ (by the induction assumption) is the certain polynomial of degree $n - 1$. As $c_n \prod_{i=1}^{n-1} \frac{c_i}{c_{i+1}} = 1$ and $P_n + p_n = \lambda \, P_{n-1}$, the claim follows.

(b) A direct computation shows that

$$(\lambda_j \,\mathrm{id}_n - \hat{C}) \, v_j = 0, \qquad (\lambda_j \,\mathrm{id}_n - \check{C}) \, w_j = 0.$$

(c) Hence $\hat{C}V = VD$, where $D = \mathrm{diag}(\lambda_1, \ldots, \lambda_n)$ is a diagonal matrix. If $\{\lambda_j\}$ are pairwise distinct then $\det V \neq 0$ and obviously $V^{-1}\hat{C}V = D$. $\square$

Consider now the infinite system of linear PDEs with functions $f_j \in C^2(\mathbb{R}^2)$

$$\partial_t \tau_i + \frac{i}{2} \sum_{j=1}^{n-1} \frac{j}{i+j-1} f_j(t,x) \, \partial_x \tau_{i+j-1} = 0, \qquad i = 1, 2, \ldots, \tag{3.14}$$

where τ_i $(i \in \mathbb{N})$ are the power sums of smooth functions $\lambda_i(t,x)$ $(1 \le i \le n)$. Let σ_j $(j \le n)$ be the elementary symmetric functions of $\{\lambda_i(t,x)\}$.

Proposition 3.2. *The matrix of the n-truncated system (3.14) (where τ_{n+i}'s are eliminated using suitable polynomials of $\tau_1, \ldots, \tau_n$, as described in Remark 1.1) is the following polynomial of the generalized companion matrix (3.13):*

$$\tilde{B} = \sum_{m=1}^{n-1} \frac{m}{2} f_m \cdot (B_{n,1})^{m-1}. \tag{3.15}$$

The eigenvalues of $\tilde{B}$ are $\tilde{\lambda}_i = \frac{1}{2} \sum_{m=1}^{n-1} m \, f_m \, \lambda_i^{m-1}$, while the corresponding eigenvectors (if λ_i's are pairwise different) read $v_i = (1, 2\lambda_i, 3\lambda_i^2, \ldots, n\lambda_i^{n-1})$.

Proof. Let $B_{n,m}$ be the matrix of the n-truncated system (3.14) with $f_j = \delta_{j,m+1}$,

$$\partial_t \tau_i + \frac{i}{2} \cdot \frac{m+1}{i+m} \partial_x \tau_{i+m} = 0, \qquad i = 1, 2, \ldots,$$

i.e., $\partial_t \vec{\tau} + B_{n,m} \partial_x \vec{\tau} = 0$. In particular, $B_{n,0} = \frac{1}{2} \mathrm{id}$. We need to prove only

$$B_{n,m} = \frac{m+1}{2} (B_{n,1})^m, \qquad 0 \le m < n, \tag{3.16}$$

therefore, $\tilde{B} = \sum_{m=1}^{n-1} f_m B_{n,m-1}$. Notice that (3.16) follows directly from the equality

$$B_{n,m} = \frac{m+1}{m} B_{n,m-1} B_{n,1}. \tag{3.17}$$

The formulae (3.17) are true for $m = 1$. We shall show that all the (i,j)-entries of the matrices $\frac{m+1}{m} B_{n,m-1} B_{n,1}$ and $B_{n,m}$ coincide.

Replacing $\partial_x \tau_{n+j}$ in (3.14) by linear combinations of $\partial_x \tau_i$ $(i \leq n)$ due to (3.19) in what follows, we find the (i,j) entry of $B_{n,m}$

$$
b_{ij}^{(n,m)} = \begin{cases} \dfrac{i(m+1)}{2(i+m)} \delta_{i+m}^{j} & \text{if } i+m \leq n, \\[3mm] (-1)^{n-j} \dfrac{i(m+1)}{2j} \beta_{n,i+m-n,j} & \text{if } i+m > n, \end{cases} \tag{3.18}
$$

for some β's (studied later in Lemma 3.3). The reader can verify using $(3.20)_1$ that for $m = 1$ the formulae (3.18) determine the matrix $B_{n,1}$ of (3.13).

Let us prove (3.17). First, assume $i + m - 1 - n \leq 0$. Then, using (3.18), we have

$$
\frac{m+1}{m} \sum_{s=1}^{n} b_{is}^{(n,m-1)} b_{sj}^{(n,1)} = \frac{m+1}{m} \sum_{s=1}^{n} \frac{im}{2(i+m-1)} \delta_{i+m-1}^{s} \frac{s}{s+1} \delta_{s}^{j-1}
$$

$$
\overset{j=s+1=i+m}{=\!=\!=} \frac{m+1}{2} \frac{i}{i+m-1} \delta_{i+m-1}^{j-1} \frac{i+m-1}{i+m}
$$

$$
= \frac{i(m+1)}{2(i+m)} \delta_{i+m}^{j}
$$

$$
= b_{ij}^{(n,m)}.
$$

Now, let $i + m - 1 - n = \tilde{m} > 0$. Then, assuming $j > 1$ and using (3.13), (3.18) and (3.20)–(3.21), we have

$$
\frac{m+1}{m} \sum_{s=1}^{n} b_{is}^{(n,m-1)} b_{sj}^{(n,1)} = \frac{m+1}{m} \left[b_{i,j-1}^{(n,m-1)} b_{j-1,j}^{(n,1)} + b_{in}^{(n,m-1)} b_{nj}^{(n,1)} \right]
$$

$$
\overset{(3.18)}{=} \frac{m+1}{m} \left((-1)^{n-j+1} \frac{im}{2(j-1)} \beta_{n,\tilde{m},j-1} \frac{j-1}{j} \right.
$$

$$
\left. + \frac{im}{2n} \beta_{n,\tilde{m},n} (-1)^{n-j} \frac{n}{j} \sigma_{n-j+1} \right)
$$

$$
\overset{(3.21)}{=} \frac{i(m+1)}{2j} (-1)^{n-j} (\beta_{n+1,\tilde{m},n+1} \sigma_{n-j+1} - \beta_{n+1,\tilde{m},j})
$$

$$
= \frac{i(m+1)}{2j} (-1)^{n-j} \beta_{n,\tilde{m}+1,j} \overset{(3.18)}{=} b_{ij}^{(n,m)}.
$$

The case $j = 1$ is similar. $\qquad\qquad\qquad\qquad\qquad\qquad\qquad\qquad\qquad\qquad\square$

Remark 3.1. The generalized companion matrix $B_{n,1}$ of (3.13) is hyperbolic if and only if λ_i's are pairwise different. In this case, by Proposition 3.2, the matrices $B_{n,m}$ $(m > 0)$ of (3.16) and the matrix $\tilde{B}$ of (3.15) are also hyperbolic.

Lemma 3.3. *The coefficients $\beta_{n,m,i}$ of the decomposition*

$$\frac{1}{n+m} \partial_x \tau_{n+m} = \sum_{i=1}^n (-1)^{n-i} \frac{1}{i} \beta_{n,m,i} \partial_x \tau_i, \qquad m > 0 \tag{3.19}$$

satisfy the following recurrence relations:

$$\beta_{n,1,i} = \sigma_{n-i+1}, \quad \beta_{n,m,i} = \beta_{n+1,m-1,n+1}\sigma_{n-i+1} - \beta_{n+1,m-1,i} \quad (m>1), \tag{3.20}$$

$$\beta_{n,m,i} = \beta_{n+j,m,i+j} \quad (1 \le i \le n, \quad m, j > 0). \tag{3.21}$$

Remark 3.2. In view of (3.21), relation (3.20)$_2$ reduces to

$$\beta_{n,m,i} = \beta_{n,m-1,n}\sigma_{n-i+1} - \beta_{n,m-1,i-1} \quad (m>1).$$

For small values of m, $m = 1,2$, we obtain from (3.19) the relations

$$\frac{1}{n+1} \partial_x \tau_{n+1} = \sum_{i=1}^n \frac{(-1)^{n-i}}{i} \sigma_{n-i+1} \partial_x \tau_i, \tag{3.22}$$

$$\frac{1}{n+2} \partial_x \tau_{n+2} = \sum_{i=1}^n \frac{(-1)^{n-i}}{i} (\sigma_1 \sigma_{n-i+1} - \sigma_{n-i+2}) \partial_x \tau_i. \tag{3.23}$$

By Proposition 3.2, the last row of $B_{n,1}$ (respectively, of $B_{n,2}$) consists of the coefficients at $\partial_x \tau_i$'s on the RHS of (3.22), (respectively, of (3.23)), and so on.

Proof (of Lemma 3.3). Let $m = 1$. The definition $\tau_i = \sum_j \lambda_j^i$ yields the equality

$$\partial_x \tau_i = i \sum_j \lambda_j^{i-1} \partial_x \lambda_j.$$

Using this, we find

$$\sum_{i=1}^n \frac{(-1)^{n-i}}{i} \sigma_{n-i+1} \partial_x \tau_i = \sum_j \left(\sum_{i=1}^n (-1)^{n-i} \sigma_{n-i+1} \lambda_j^{i-1} \right) \partial_x \lambda_j.$$

Define the polynomial $P_n(x) = \lambda^n - \sigma_1(x)\lambda^{n-1} + \ldots + (-1)^n \sigma_n(x)$. As $\lambda_j(x)$ are the roots of P_n, we obtain the identity

$$\frac{1}{n+1} \partial_x \tau_{n+1} - \sum_{i=1}^n \frac{(-1)^{n-i}}{i} \sigma_{n-i+1} \partial_x \tau_i$$

$$= \sum_j \left(\lambda_j^n - \sigma_1 \lambda_j^{n-1} + \ldots + (-1)^n \sigma_n \right) \partial_x \lambda_j = 0$$

that proves (3.22). Hence, $\beta_{n,1,i} = \sigma_{n-i+1}$.

In order to prove the recurrence relation in (3.20), assume temporarily that $\lambda_{n+1} = \varepsilon$, $\tilde{n} = n+1$, and $\tilde{m} = m-1$. Hence,

$$\frac{1}{n+m} \partial_x \tau_{n+m} = \frac{1}{\tilde{n}+\tilde{m}} \partial_x \tau_{\tilde{n}+\tilde{m}} \big|_{\varepsilon=0}.$$

Then, we may put $\varepsilon = 0$ and replace $\partial_x \tau_{n+1}(x)$ via (3.22) to obtain

$$\frac{1}{\tilde{n}+\tilde{m}}\partial_x \tau_{\tilde{n}+\tilde{m}} = \sum_{i=1}^{\tilde{n}} \frac{(-1)^{\tilde{n}-i}}{i} \beta_{\tilde{n},\tilde{m},i}\, \partial_x \tau_i$$

$$= \beta_{n+1,m-1,n+1}\frac{1}{n+1}\partial_x \tau_{n+1} + \sum_{i=1}^{n} \frac{(-1)^{n-i+1}}{i}\beta_{n+1,m-1,i}\, \partial_x \tau_i$$

$$\overset{\varepsilon=0}{=} \sum_{i=1}^{n} \frac{(-1)^{n-i}}{i}\left(\beta_{n+1,m-1,n+1}\, \sigma_{n-i+1} - \beta_{n+1,m-1,i}\right)\partial_x \tau_i$$

that completes the proof of (3.20). For $m = 2$, we deduce from (3.20) the equality

$$\beta_{n,2,i} = \beta_{n+1,1,n+1}\, \sigma_{n-i+1} - \beta_{n+1,1,i} = \sigma_1 \sigma_{n-i+1} - \sigma_{n-i+2}$$

which proves (3.23).

Finally, we prove (3.21) by induction on m. For $m = 1$, using $(3.20)_1$, we have

$$\beta_{n+j,1,i+j} = \sigma_{(n+j)-(i+j)+1} = \sigma_{n-i+1} = \beta_{n,1,i}.$$

Assuming (3.21) for $m - 1$ and using $(3.20)_2$, we deduce it for m:

$$\beta_{n+j,m,i+j} = \beta_{(n+j)+1,m-1,(n+j)+1}\sigma_{(n+j)-(i+j)+1} - \beta_{(n+j)+1,m-1,i+j}$$

$$= \beta_{n+1,m-1,n+1}\sigma_{n-i+1} - \beta_{n+1,m-1,i} = \beta_{n,m,i}.$$

This completes the proof of (3.21). □

Example 3.3. For $f_j = \delta_{j1}$, (3.14) reduces to the linear system

$$\partial_t \tau_i + \frac{1}{2}\partial_x \tau_i = 0,$$

whose solution is a simple wave along the x-axis: $\tau_i = \tau_i^0(t - 2x)$.

Consider the slightly more complicated cases.

1. For $f_j = \delta_{j2}$, (3.14) reduces to the system

$$\partial_t \tau_i + \frac{i}{i+1}\partial_x \tau_{i+1} = 0, \quad i = 1, 2, \dots \tag{3.24}$$

The n-truncated system (3.24) reads: $\partial_t \vec{\tau} + B_{n,1}\partial_x \vec{\tau} = 0$.

For $n = 2$, using (3.22), we have just two PDEs

$$\partial_t \tau_1 = -\frac{1}{2}\partial_x \tau_2, \quad \partial_t \tau_2 = -\frac{2}{3}\partial_x \tau_3 = (\tau_1^2 - \tau_2)\partial_x \tau_1 - \tau_1 \partial_x \tau_2.$$

The matrix $B_{2,1} = \begin{pmatrix} 0 & \frac{1}{2} \\ -2\sigma_2 & \sigma_1 \end{pmatrix}$ has the characteristic polynomial $P_2 = \lambda^2 - \sigma_1\lambda + \sigma_2$. If the roots λ_1 and λ_2 of P_2 are distinct, the eigenvectors of $B_{2,1}$ are equal to $v_j = (1, 2\lambda_j)$, $j = 1, 2$. If $\lambda_1 = \lambda_2 \neq 0$ then $B_{2,1}$ has one eigenvector only, hence the system (3.24) is not hyperbolic in the t-direction.

For $n = 3$, (3.24) reduces to the quasilinear system of three PDEs with the matrix

$$B_{3,1} = \begin{pmatrix} 0 & \frac{1}{2} & 0 \\ 0 & 0 & \frac{2}{3} \\ 3\sigma_3 & -\frac{3}{2}\sigma_2 & \sigma_1 \end{pmatrix}.$$

The characteristic polynomial of $B_{3,1}$ is $P_3 = \lambda^3 - \sigma_1\lambda^2 + \sigma_2\lambda - \sigma_3$, the eigenvalues are λ_j, and the eigenvectors are $v_j = (1, 2\lambda_j, 3\lambda_j^2)$, $j = 1, 2, 3$.

2. For $f_j = \delta_{j3}$, (3.14) reduces to the system

$$\partial_t \tau_i + \frac{3i}{2(i+2)}\partial_x \tau_{i+2} = 0, \qquad i = 1, 2, \ldots \tag{3.25}$$

The matrix of this n-truncated system is $B_{n,2} = \frac{3}{2}(B_{n,1})^2$.

For $n = 3$, (3.25) reduces to the system of three quasilinear PDEs

$$\partial_t \tau_1 = -\frac{1}{2}\partial_x \tau_3, \quad \partial_t \tau_2 = -\frac{3}{4}\partial_x \tau_4, \quad \partial_t \tau_3 = -\frac{9}{10}\partial_x \tau_5,$$

where $\partial_x \tau_4$ and $\partial_x \tau_5$ should be expressed using (3.22) and (3.23). Hence, the matrix of this system is

$$B_{3,2} = \frac{3}{2}(B_{3,1})^2 = \begin{pmatrix} 0 & 0 & \frac{1}{2} \\ 3\sigma_3 & -\frac{3}{2}\sigma_2 & \sigma_1 \\ \frac{9}{2}\sigma_1\sigma_3 & \frac{9}{4}(\sigma_3 - \sigma_1\sigma_2) & \frac{3}{2}(\sigma_1^2 - \sigma_2) \end{pmatrix}.$$

Its eigenvalues are $\frac{3}{2}\lambda_j^2$, and the eigenvectors are the same as for $B_{3,1}$.

Similarly, for $n = 4$, (3.25) reduces to the quasilinear system with the matrix

$$B_{4,2} = \frac{3}{2}(B_{4,1})^2 = \begin{pmatrix} 0 & 0 & \frac{1}{2} & 0 \\ 0 & 0 & 0 & \frac{3}{4} \\ -\frac{9}{2}\sigma_4 & \frac{9}{4}\sigma_3 & -\frac{3}{2}\sigma_2 & \frac{9}{8}\sigma_1 \\ -6\sigma_1\sigma_4 & 3(\sigma_1\sigma_3 - \sigma_4) & 2(\sigma_3 - \sigma_1\sigma_2) & \frac{3}{2}(\sigma_1^2 - \sigma_2) \end{pmatrix}$$

with eigenvalues $\tilde{\lambda}_j = \frac{3}{2}\lambda_j^2$, and eigenvectors $v_j = (1, 2\lambda_j, 3\lambda_j^2, 4\lambda_j^3)$, $j = 1, 2, 3, 4$.

3. For $f_j = \delta_{j4}$, (3.14) reduces to the system

$$\partial_t \tau_i + \frac{2i}{i+3}\partial_x \tau_{i+3} = 0, \qquad i = 1, 2, \ldots$$

with the corresponding matrix $B_{n,3} = 2(B_{n,1})^3$. For example,

$$B_{4,3} = \begin{pmatrix} 0 & 0 & 0 & \frac{1}{2} \\ -4\sigma_4 & 2\sigma_3 & -\frac{4}{3}\sigma_2 & \sigma_1 \\ -6\sigma_1\sigma_4 & 3(\sigma_1\sigma_3 - \sigma_4) & 2(\sigma_3 - \sigma_1\sigma_2) & \frac{3}{2}(\sigma_1^2 - \sigma_2) \\ 8\sigma_4(\sigma_2 - \sigma_1^2) & 4(\sigma_1^2\sigma_3 - \sigma_2\sigma_3 - \sigma_1\sigma_4) & \frac{8}{3}(\sigma_2^2 - \sigma_4 + \sigma_1\sigma_3 - \sigma_1^2\sigma_2) & 2(\sigma_3 - 2\sigma_1\sigma_2 + \sigma_1^3) \end{pmatrix}$$

has the eigenvalues $2\lambda_j^3$. If $\lambda_i < \lambda_j$ $(i < j)$, the four linearly independent eigenvectors of $B_{4,3}$ are $\mathbf{v}_j = (1, 2\lambda_j, 3\lambda_j^2, 4\lambda_j^3)$, $j \leq 4$.

This series of examples can be continued as long as one desires.

3.4 Existence and Uniqueness Results (Main Theorems)

We now formulate our results concerning existence/uniqueness of EGFs and their corollaries. For brevity, we shall omit the index t for t-dependent tensors $A, b, \hat{b}_j$ and functions τ_i, σ_i. The following theorem concerns an EGF of type (a) and is essential in the proof of Theorem 3.2 (for an EGF of type (b)).

Theorem 3.1. *Let (M, g_0) be a closed Riemannian manifold with a codimension-one foliation $\mathscr{F}$. Given functions $\hat{f}_j \in C^2(M \times \mathbb{R})$, there exists a unique smooth solution g_t of (3.1) of type (a) defined on some positive time interval $[0, \varepsilon)$.*

In particular, there exists a unique smooth solution g_t, $t \in [0, \varepsilon)$ to the PDEs

$$\partial_t g_t = \sum_{j=0}^{n-1} a_j(t)\hat{b}_j, \quad a_j \in C^2(\mathbb{R}).$$

By Proposition 3.2, the n-truncated system (3.2) (where τ_{n+i}'s are replaced by suitable polynomials of $\tau_1, \ldots, \tau_n$) has the form $\partial_t \vec{\tau} = \widehat{B}_t \partial_x \vec{\tau}$ with the $n \times n$ matrix $\widehat{B}_t = \tilde{A} + \tilde{B}$, where

$$\tilde{A} = (\tilde{A}_{ij}), \quad \tilde{A}_{ij} = \frac{i}{2}\sum_{m=0}^{n-1} \tau_{i+m-1} f_{m,\tau_j}, \quad \tilde{B} = \sum_{m=1}^{n-1} \frac{m}{2} f_m \cdot (B_{n,1})^{m-1}, \quad (3.26)$$

and $B_{n,1}$ is the *generalized companion matrix* to the characteristic polynomial of A_N.

Example 3.4. (a) If g_t satisfy $\partial_t g_t = F(\vec{\tau}, t)\hat{g}_t$ (i.e., $f_j = \delta_{j0}F$) then, by (3.26),

$$\tilde{B} = 0, \quad \tilde{A} = (\tilde{a}_{ij}), \quad \text{where} \quad \tilde{a}_{ij} = \frac{i}{2}\tau_{i-1}F_{,\tau_j}(\vec{\tau}, 0).$$

The system (3.2) reduces to

$$\partial_t \tau_i + \frac{i}{2}\tau_{i-1}\sum_{j=1}^{n} F_{,\tau_j}(\vec{\tau}, t)N(\tau_j) = 0, \quad i = 1, 2, \ldots \quad (3.27)$$

The matrix $\tilde{A}$ of n-truncated system (3.27) is *hyperbolic* if for any $q \in M$

$$\text{either } 2\,\text{Tr}\,\tilde{A} = \sum_i i\,\tau_{i-1}F_{,\tau_i} \neq 0 \text{ or } \tilde{A} \equiv 0 \text{ on the } N\text{-curve through q.} \qquad (H_1)$$

(b) If the g_t satisfy $\partial_t g_t = F(\overrightarrow{\tau},t)\hat{b}_1$ (i.e., $f_j = \delta_{j1}F$) then, again by (3.26),

$$\tilde{B} = \frac{1}{2}F(\overrightarrow{\tau},0)\,\widehat{\text{id}}, \quad \tilde{A} = (\tilde{a}_{ij}), \quad \text{where} \quad \tilde{a}_{ij} = \frac{i}{2}\,\tau_i F_{,\tau_j}(\overrightarrow{\tau},0).$$

The system (3.2) reduces to

$$\partial_t \tau_i + \frac{1}{2}F(\overrightarrow{\tau},t)N(\tau_i) + \frac{i}{2}\,\tau_i \sum_{j=1}^{n} F_{,\tau_j}(\overrightarrow{\tau},t)N(\tau_j) = 0, \quad i = 1,2,\dots \quad (3.28)$$

The matrix $\tilde{A} + \tilde{B}$ of the n-truncated system (3.28) is *hyperbolic* if for any $q \in M$

$$\text{either } 2\,\text{Tr}\,\tilde{A} = \sum_i i\,\tau_i F_{,\tau_i} \neq 0 \text{ or } \tilde{A} \equiv 0 \text{ along the } N\text{-curve through } q. \qquad (H_2)$$

Recall that an n-by-n matrix is *hyperbolic* (see Sect. 3.3.2) if its right eigenvectors are real and span $\mathbb{R}^n$.

The central result here is the following.

Theorem 3.2 (Short time existence). *Let* (M,g_0) *be a closed Riemannian manifold with a foliation* $\mathscr{F}$ *and a unit normal N. If the matrices* $\tilde{A} + \tilde{B}$ *and* $\tilde{B}$ *of (3.26) are hyperbolic for all* $q \in M$ *and* $t = 0$ *then the EGF (3.1) of type (b) has a unique smooth solution* g_t *defined on some positive time interval* $[0,\varepsilon)$.

The proof of Theorem 3.1 (see Sect. 3.5.2) follows methods of the theory of first-order hyperbolic PDEs with one space variable.

The proof of Theorem 3.2 (see Sect. 3.5.2) consists of the following steps:

(1) Power sums τ_i are recovered on M (as a unique solution to a quasilinear hyperbolic system of PDEs) for some positive time interval $[0,\varepsilon)$, see Lemma 3.5 and Proposition 3.2.
(2) Given (τ_i) (of Step 1), the metric g_t is recovered on M (as a unique solution to certain quasilinear system of PDEs), see Theorem 3.1.
(3) The τ_i-s of the g_t-principal curvatures of $\mathscr{F}$ (of Step 2) are shown to coincide with τ_i (of Step 1), see Theorem 3.1 and Lemma 3.6.

Remark that the solution in Theorem 3.2 is unique if only $\tilde{A} + \tilde{B}$ is hyperbolic. For $f_j = 0$ ($j \geq 2$), Theorem 3.2 holds under the weaker condition that only the matrix $\tilde{A}$ is hyperbolic for all $q \in M$ and $t = 0$.

In what follows we denote by $\mathscr{L}_Z$ the Lie derivative along a vector field Z.

Corollary 3.1. *Let* (M,g_0) *be a closed Riemannian manifold with a codimension-one foliation* $\mathscr{F}$ *and unit normal N. If* $F \in C^2(\mathbb{R}^{n+1})$ *and the condition* (H_2) *is satisfied at* $t = 0$ *and any* $q \in M$ *then there is a unique smooth solution to the EGF*

$$\partial_t g_t = F(\overrightarrow{\tau},t)\hat{b}_1, \quad t \in [0,\varepsilon) \qquad (3.29)$$

for some $\varepsilon > 0$. Furthermore, g_t can be determined from the system $\mathscr{L}_{Z_t} g_t = 0$ with $Z_t = \partial_t + \frac{1}{2}F(\vec{\tau},t)N$, where $\vec{\tau}$ are the unique smooth solution to (3.28).

Corollary 3.2. *Let (M, g_0) be a closed Riemannian manifold with a codimension-one foliation $\mathscr{F}$ and a unit normal N. If $F \in C^2(\mathbb{R}^{n+1})$ and the condition (H_1) is satisfied at $t = 0$ and any $q \in M$ then there is a unique smooth solution to the EGF*

$$\partial_t g_t = F(\vec{\tau},t)\,\hat{g}_t, \quad t \in [0,\varepsilon) \tag{3.30}$$

for some $\varepsilon > 0$. Furthermore, $\hat{g}_t = \hat{g}_0 \exp(\int_0^t F(\vec{\tau},t)\,dt)$, where the power sums $\vec{\tau}$ are the unique solution to (3.27).

Example 3.5. For $f_1 = c = const$ and $f_j = 0$ $(j \neq 1)$, i.e., for the EGF with $h(b) = C\hat{b}_1$, $C \in \mathbb{R}$, the system (3.2) (see also (3.28) for $f = c$) reduces to the linear PDE

$$\partial_t \tau_i + (c/2)N(\tau_i) = 0.$$

The above PDE, $\partial_t \vec{\tau} = \tilde{B}_t \, \partial_x \vec{\tau}$, can be interpreted on $M \times \mathbb{R}$ by saying that τ_i are constant along the orbits of the vector field $X = \partial_t + (C/2)N$.

If (ψ_t) denotes the flow of $(c/2)N$ on M then the flow (ϕ_t) of X is given by:

$$\phi_t(p,s) = (\psi_t(p),\ t+s) \quad \text{for} \quad q \in M,\ s \in \mathbb{R},$$

therefore, ϕ_t maps the level surface $M_s = M \times \{s\}$ onto M_{t+s}, in particular, M_0 onto M_t.

This example implies

Corollary 3.3. *If $f_1 = const$ and $f_j = 0$ for all $j \neq 1$ (for the EGF) then for all i and t, the following equality holds:*

$$\tau_i^t = \tau_i^0 \circ \phi_{-t}.$$

In particular, if $\tau_i^0 = const$ for some i then $\tau_i = const$ for all t.

A closed manifold M equipped with a foliation $\mathscr{F}$ admits a Riemannian structure g for which all the leaves are minimal ($\tau_1 = 0$ in our terminology) if and only if $\mathscr{F}$ is *topologically taut*, that is every leaf meets a loop transverse to the foliation [49]. Known proofs of existence of such metrics use the Hahn–Banach Theorem and are not constructive. The above observations show how to produce a 1-parameter family of metrics with $\tau_{2j+1} = 0$ (with fixed j) starting from one of such metrics.

3.5 The General Case

Here we prove the results of Sect. 3.4 about EGFs. First, in Sect. 3.5.1, we solve PDEs for τ's. Then, in Sect. 3.5.2, we use this to prove local existence and uniqueness of EGFs. Section 3.5.3 is devoted to proofs of corollaries.

3.5.1 Searching for Power Sums

Let g_t satisfy (3.1) and N_t be the g_t-unit normal vector field to $\mathscr{F}$ on M.

It is easy to see that $\partial_t N_t = 0$, therefore $N_t = N$ for all t, where N is the unit normal of $\mathscr{F}$ on (M, g_0). In fact, for any vector field X tangent to $\mathscr{F}$ one has

$$0 = \partial_t g_t(X, N_t) = h(b_t)(X, N_t) + g_t(X, \partial_t N_t) = g_t(X, \partial_t N_t),$$

and similarly

$$0 = \partial_t g_t(N_t, N_t) = h(b_t)(N_t, N_t) + 2g_t(N_t, \partial_t N_t) = 2g_t(N_t, \partial_t N_t).$$

Now let ∇^t be the Levi-Civita connection on (M, g_t). Then $\Pi_t = \partial_t \nabla^t$ is a $(1,2)$-tensor field on M. Following (2.11), we write for all t-independent vector fields X, Y, and Z

$$g_t(\Pi_t(X, Y), Z) = \frac{1}{2}\left((\nabla_X^t h_t)(Y, Z) + (\nabla_Y^t h_t)(X, Z) - (\nabla_Z^t h_t)(X, Y)\right). \quad (3.31)$$

The next lemma is a consequence of Lemma 2.9; we shall prove it directly.

Lemma 3.4. *For an EGF of type (b), on the tangent bundle of $\mathscr{F}$ we have*

$$\partial_t b(X, Y) = h_t(AX, Y) - \frac{1}{2}\sum_{j=0}^{n-1}\left(N(f_j)\,g_t(A^j X, Y) + f_j\,g_t(\nabla_N^t(A^j)X, Y)\right), \quad (3.32)$$

$$\partial_t A = -\frac{1}{2}\sum_{j=0}^{n-1}\left(N(f_j)A^j + f_j\nabla_N^t(A^j)\right) = -\frac{1}{2}\nabla_N^t h(A). \quad (3.33)$$

Proof. By definition, $b(X, Y) = g_t(\nabla_X^t Y, N)$, and $h(\cdot, N) = 0$. Using (3.31) and the identity $h(AX, Y) = h(AY, X)$, we obtain

$$\begin{aligned}
\partial_t b(X, Y) &= \partial_t g_t(\nabla_X^t Y, N) = (\partial_t g_t)(\nabla_X^t Y, N) + g_t(\partial_t \nabla_X^t Y, N) \\
&= \frac{1}{2}\left((\nabla_X^t h_t)(Y, N) + (\nabla_Y^t h_t)(X, N) - (\nabla_N^t h_t)(X, Y)\right) + h_t(\nabla_X^t Y, N) \\
&= \frac{1}{2}\left(h_t(AX, Y) + h_t(AY, X) - N(h_t(X, Y))\right) \\
&= h_t(AX, Y) - \frac{1}{2}\sum_{j=0}^{n-1}\left(f_j\,g_t(\nabla_N^t(A^j)X, Y) + N(f_j)\,g_t(A^j X, Y)\right).
\end{aligned}$$

This proves (3.32). Now, (3.33) follows from (3.32) and

$$\begin{aligned}
g_t((\partial_t A)X, Y) &= g_t(\partial_t(AX), Y) = \partial_t(g_t(AX, Y)) - (\partial_t g_t)(AX, Y) \\
&= \partial_t b(X, Y) - h_t(AX, Y). \qquad \square
\end{aligned}$$

Remark 3.3. We prove (3.2) directly, applying iA^{i-1} to both sides of (3.33) we obtain the PDE

$$iA^{i-1}\partial_t A = -\frac{i}{2}\sum_{j=0}^{n-1}\left(N(f_j)A^{i+j-1}+f_j A^{i-1}\nabla_N^t A^j\right).$$

Taking the trace of both sides of the above equality, and using the identities

$$\partial_t \tau_i = \mathrm{Tr}\left(\partial_t A^i\right) = i\,\mathrm{Tr}\left(A^{i-1}\partial_t A\right) \tag{3.34}$$

and (3.4) for $i, j > 0$, we obtain (3.2).

From Proposition 3.2 it follows directly that the functions $\vec{\tau} = (\tau_1,\ldots,\tau_n)$ satisfy the system of n quasilinear PDEs, whose matrix can be built using a generalized companion matrix of the characteristic polynomial of A.

Lemma 3.5. *The n-truncated system (3.2) has the form*

$$\partial_t \vec{\tau} + (\tilde{A}+\tilde{B})\partial_x \vec{\tau} = 0$$

with $\tilde{A}$ and $\tilde{B}$ given by (3.26).

The next lemma deals with the evolution equation for an EGF of type (a).

Lemma 3.6. *Let g_t be the solution to the EGF (3.1) of type (a). Then the Weingarten operator A of $\mathscr{F}$ with respect to g_t satisfies*

$$\partial_t A = -\frac{1}{2}\left(N(\tilde{f}_0)\,\widehat{\mathrm{id}} + \sum_{j=1}^{n-1}\left(N(\tilde{f}_j)A^j + \tilde{f}_j \cdot \nabla_N^t(A^j)\right)\right), \tag{3.35}$$

and τ_i $(i \geq 1)$ (the power sums of the eigenvalues of A) satisfy the PDEs

$$\partial_t \tau_i = -\frac{i}{2}\left\{\tau_{i-1}N(\tilde{f}_0) + \sum_{j=1}^{n-1}\left(\frac{j}{i+j-1}\tilde{f}_j N(\tau_{i+j-1}) + \tau_{i+j-1}N(\tilde{f}_j)\right)\right\}. \tag{3.36}$$

The n-truncated system (3.36) has the form

$$\partial_t \vec{\tau} + \left(\sum_{j=1}^{n-1} j\tilde{f}_j\,(B_{n,1})^{j-1}\right)N(\vec{\tau}) = a,$$

where $B_{n,1}$ is the generalized companion matrix (3.13), $a = (a_1,\ldots,a_n)$, and

$$a_i = -\frac{i}{2}\sum_{j=0}^{n-1}N(\tilde{f}_j)\,\tau_{i+j-1} \quad (1 \leq i \leq n).$$

Proof. The proof of (3.35) is similar to that of (3.33). On $T\mathscr{F}$ we obtain

$$g_t((\partial_t A)X,Y) = -\frac{1}{2}N(h_t(X,Y))$$

$$= -\frac{1}{2}\left(N(\tilde{f}_0)g_t(X,Y) + \sum_{j=1}^{n-1}\left(N(\tilde{f}_j)g_t(A^jX,Y)\right.\right.$$

$$\left.\left. +\tilde{f}_j\,g_t(\nabla_N^t(A^j)X,Y)\right)\right).$$

Following the lines of the proof of (3.2) in Lemma 3.5, we can deduce from the above our system (3.36). By Proposition 3.2, the n-truncated system (3.36) has the required form.   □

Lemmas 3.5 and 3.6 together with Theorem A provide existence and uniqueness results for the symmetric functions $\overrightarrow{\tau}^t$ satisfying conditions following from (3.1). In particular, this allows us to reduce the existence and uniqueness problems for EGFs of type (b) to those for type (a), as we do in the proof of Theorem 3.2 in the next section.

3.5.2   Local Existence of Metrics (Proofs of the Main Theorems)

Given a Riemannian metric g on a foliated manifold $(M,\mathscr{F})$, the symmetric tensor $h(b)$ defined by (3.1) can be expressed in terms of the first partial derivatives of g. Therefore, $g \mapsto h(b)$ is a first-order partial differential operator. For any $\mathscr{F}$-truncated symmetric $(0,2)$-tensor S the equation

$$h(b) = S \tag{3.37}$$

has the form of a nonlinear system of first-order PDEs. A particular case of (3.37) is the Einstein type relation $h(b) = \rho\,\hat{g}$ for some function (or constant) ρ on M.

There are several obvious obstructions to the existence of solutions to (3.37) even at a single point. For example, if $h(b) = f_{2j}\hat{b}_{2j}$ (for some integer j and a function f_{2j}) and S is neither positive nor negative definite then (3.37) has no solutions at q.

Proof (of Theorem 3.1). Take biregular foliated coordinates $(x_0, x_1, \ldots x_n)$ on $U_q \subset M$ (with center at q); see Lemma 2.1, and the metric (2.2). Then, $N = \partial_0/\sqrt{g_{00}}$ is the unit normal to $\mathscr{F}$. Set $\psi_{ab} = g_{ab,0}$. The system (3.1) (for f_j of type (a)) along a trajectory $\gamma : x \mapsto \gamma(x)$ of ∂_0 has the form

$$\partial_t g_{ij} = F_{ij}(g_{ab}, \psi_{ab}, t, x), \tag{3.38}$$

where $F_{ij} := h(b)_{ij}$. In view of symmetry, $\psi_{ab} = \psi_{ba}$ and $F_{ij} = F_{ji}$, we shall assume $1 \le i \le j \le n$ and $1 \le a \le b \le n$. For example, if $f_m = 0$ $(m > 1)$ then (3.38) is the hyperbolic (diagonal) system

$$\partial_t g_{ij} = f_0(q,t)g_{ij} - \frac{1}{2}g_{00}^{-1/2}f_1(q,t)\psi_{ij},$$

that completes the proof in this case.

Now let $f_m \neq 0$ for some $m > 1$ (e.g., general f_m). We may assume that $A\partial_j = k_j \partial_j$, $g(\partial_i, \partial_j) = \delta_{ij}$ $(i,j > 0)$ and $g_{00} = 1$ at the point q for $t = 0$. (By Lemma 2.2, we have $(b_m)_{ij} = (-1/2)^m \psi_{ij}^m \delta_{ij}$ at q for $t = 0$).

Differentiating (3.38) with respect to x and t, we obtain

$$\partial_0 p_{ij} = \partial_0 F_{ij} + \sum_{a,b} \left[\frac{\partial F_{ij}}{\partial g_{ab}} \partial_0 g_{ab} + \frac{\partial F_{ij}}{\partial \psi_{ab}} \partial_0 \psi_{ab} \right],$$

$$\partial_t p_{ij} = \partial_t F_{ij} + \sum_{a,b} \left[\frac{\partial F_{ij}}{\partial g_{ab}} \partial_t g_{ab} + \frac{\partial F_{ij}}{\partial \psi_{ab}} \partial_t \psi_{ab} \right], \tag{3.39}$$

where

$$p_{ij} := \partial_t g_{ij}, \quad \psi_{ij} := g_{ij,0}, \quad F_{ij} := h(b)_{ij}.$$

Since g is of class C^2, we conclude that

$$\partial_t \psi_{ab} = \frac{\partial^2 g_{ab}}{\partial t \partial x_0} = \partial_0 p_{ab}.$$

Hence (3.39) together with (3.38) may be written in the form

$$\partial_t g_{ij} = F_{ij}(\{g_{ab}\}, \{\psi_{ab}\}, t, x),$$

$$\partial_t \psi_{ij} - \sum_{a,b} \frac{\partial F_{ij}}{\partial \psi_{ab}} \partial_0 \psi_{ab} = \partial_0 F_{ij} + \sum_{a,b} \frac{\partial F_{ij}}{\partial g_{ab}} \psi_{ab},$$

$$\partial_t p_{ij} - \sum_{a,b} \frac{\partial F_{ij}}{\partial \psi_{ab}} \partial_0 p_{ab} = \partial_t F_{ij} + \sum_{a,b} \frac{\partial F_{ij}}{\partial g_{ab}} p_{ab}. \tag{3.40}$$

The above quasilinear system consists of parts:

(i)   our original equation $(3.40)_1$,
(ii) the corresponding equation $(3.40)_2$ for $\partial_t A$, and
(iii) the equation $(3.40)_3$ for $\partial_t^2 g$ following from the previous ones.

In general, the following $\frac{1}{2}n(n+1) \times \frac{1}{2}n(n+1)$ matrix is not symmetric:

$$d_\psi F = \left\{ \frac{\partial F_{ij}}{\partial \psi_{ab}} \right\}, \quad i \leq j, \ a \leq b.$$

We claim that it is hyperbolic. If we change the local coordinate system on M, the components F_{ij} $(i \leq j)$ and ψ_{ab} $(a \leq b)$ at q will be transformed by the same tensor low. Notice that the above $d_\psi F$ is a $(1,1)$-tensor on the vector bundle of symmetric

$(0,2)$-tensors on $T\mathscr{F}$. Hence, $d_\psi F(q)$ can be seen as the linear endomorphism of the space of symmetric $(0,2)$-tensors on $T_q\mathscr{F}$.

The hyperbolicity is a pointwise property, so can be considered at any point $q \in M$ in a special bifoliated chart around q, for example, such that $g_{ij} = \delta_{ij}$ and $A_{ij}(q) = k_i\delta_{ij}$ at q. (Indeed, (k_i) are the principal curvatures of $\mathscr{F}$ at q for $t = 0$). In this chart, our calculations show that the matrix $d_\psi F$ is diagonal, so has real eigenvalues (vectors) at q. Indeed, for $t = 0$ one may find at the point q:

$$\frac{\partial F_{ij}}{\partial \psi_{ab}} = \sum_{m \geq 1} f_m(q)\,\mu(m)_{ij}\,\delta^{\{i,j\}}_{\{a,b\}}, \qquad \mu(m)_{ij} = \sum_{\alpha+\beta=m-1} k_i^\alpha k_j^\beta.$$

The order of indices of $d_\psi F$ is $[1,1],[1,2],\dots,[1,n],[2,2],[2,3],\dots,[2,n],\dots,[n,n]$. For example, for $F = b_2$ (i.e., $f_j = \delta_{j2}$) the above matrix in an orthonormal frame at any point is

$$\frac{\partial(b_2)_{ij}}{\partial \psi_{ab}} = \frac{1}{4}\begin{bmatrix} 2\psi_{11} & 2\psi_{12} & 2\psi_{13} & 0 & 0 & 0 \\ \psi_{12} & \psi_{11}+\psi_{22} & \psi_{23} & \psi_{12} & \psi_{13} & 0 \\ \psi_{13} & \psi_{23} & \psi_{11}+\psi_{33} & 0 & \psi_{12} & \psi_{13} \\ 0 & 2\psi_{12} & 0 & 2\psi_{22} & 2\psi_{23} & 0 \\ 0 & \psi_{13} & \psi_{12} & \psi_{23} & \psi_{22}+\psi_{33} & \psi_{23} \\ 0 & 0 & 2\psi_{13} & 0 & 2\psi_{23} & 2\psi_{33} \end{bmatrix}$$

with the order of indices $[1,1],[1,2],[1,3],[2,2],[2,3],[3,3]$. At q for $t = 0$ (i.e., $\psi_{ab} = 0$, $a \neq b$ and $\psi_{aa} = k_a$) it is diagonal with the elements $\mu(2)_{ab} = k_a + k_b$. Let $A_0 = [\frac{1}{2},1,1,\frac{1}{2},1,\frac{1}{2}]$ be the diagonal matrix. Then the matrix $A_1 = A_0\,d_\psi(b_2)$ is symmetric, and our system $(3.40)_2$ for $h(b) = b_2$ is "symmetrizable": $A_0\partial_t\psi - A_1\partial_0\psi = \{\text{free terms}\}$.

Therefore, (3.40) for the functions $g_{ij}(t,x)$, $p_{ij}(t,x)$ and $\psi_{ij}(t,x)$ with $i \leq j$ is the hyperbolic system which is "symmetrizable" in the sense of [51, p. 370]. Indeed, multiplying n columns (corresponding to $i = j$) of the matrix $d_\psi F$ in an orthonormal frame by the factor $\frac{1}{2}$, we obtain the symmetric matrix. By Theorem A (in Section 3.3.2), given $q \in M$ there exists a unique solution to (3.40) which is defined in U_q along the N-curve through q for some time interval $[0, \varepsilon_q)$ and satisfies the initial conditions

$$g_{ij}(0,x) = (g_0)_{ij}(x), \quad p_{ij}(0,x) = h(b_0)_{ij}(x), \quad \psi_{ij}(0,x) = (\partial_0 g)_{ij}(x).$$

Again by Theorem A (see definitions of K and $\bar\Omega$), the value ε_q continuously depends on $q \in M$. The claim follows from the above and compactness of M. □

Proof (of Theorem 3.2). Let A_0 and $\vec{\tau}^{\,0}$ be the values of extended Weingarten operator and power sums of the principal curvatures k_i of (the leaves of) $\mathscr{F}$ determined on $(M,\mathscr{F})$ by a given metric g_0:

(a) *Uniqueness.* Assume that $g_t^{(1)}, g_t^{(2)}$ are two solutions to (3.1) with the same initial metric g_0. Functions $\vec{\tau}^{\,t,1}, \vec{\tau}^{\,t,2}$, corresponding to $g_t^{(1)}, g_t^{(2)}$, satisfy (3.2)

and have the same initial value $\overrightarrow{\tau}^0$. By Lemma 3.5 and Theorem A, $\overrightarrow{\tau}^{t,1} = \overrightarrow{\tau}_t^{t,2} = \overrightarrow{\tau}^t$ on some positive time interval $[0,\varepsilon_1)$. Hence $g_t^{(1)}, g_t^{(2)}$ satisfy (3.1) of type (a) with known coefficients $\tilde{f}_j(p,t) := f_j(\overrightarrow{\tau}^t(p),t)$. By Theorem 3.1, $g_t^{(1)} = g_t^{(2)}$ on some positive time interval $[0,\varepsilon_2)$.

(b) *Existence.* By Lemma 3.5 and Theorem A, there is a unique solution $\overrightarrow{\tau}^t$ to (3.2) on some positive time interval $[0,\varepsilon^*)$. By Theorem 3.1, the EGF (3.1) of type (a) with known functions $\tilde{f}_j(\cdot,t) := f_j(\overrightarrow{\tau}^t,t)$ has a unique solution g_t^* ($g_0^* = g_0$) for $0 \leq t < \varepsilon^*$. The Weingarten operator A_t^* ($A_0^* = A_0$) of $(M,\mathscr{F},g_t^*)$ satisfies (3.35), hence the power sums of its eigenvalues, $\overrightarrow{\tau}^{t,*}$ ($\overrightarrow{\tau}^{0,*} = \overrightarrow{\tau}^0$), satisfy (3.36) with the same coefficient functions $\tilde{f}_j$. By Lemma 3.6 and Theorem A, the solution of this problem is unique, hence $\overrightarrow{\tau}^t = \overrightarrow{\tau}^{t,*}$, i.e., $\overrightarrow{\tau}^t$ are power sums of eigenvalues of A_t^*. Finally, g_t^* is a solution to (3.1) such that $\overrightarrow{\tau}^t$ are power sums of the principal curvatures of the leaves in this metric. □

3.5.3 Proofs of the Corollaries

In all the proofs given later, q is a point of M, $\gamma : x \mapsto \gamma(x)$ ($\gamma(0) = q$, $x \in \mathbb{R}$) is the N-curve, and N is the unit normal of $\mathscr{F}$.

Proof (of Corollary 3.1). By Lemma 3.5, we have (3.2), which in our case reduces to the system (3.28). Using this system, we build the initial value problem in the (x,t)-plane for the vector function $\overrightarrow{\tau}(x,t) = \overrightarrow{\tau}(\gamma(x),t)$

$$\partial_t \overrightarrow{\tau} + \left(\frac{1}{2}F(\overrightarrow{\tau},t)\,\mathrm{id}_n + \tilde{A}(\overrightarrow{\tau},t)\right)\partial_x \overrightarrow{\tau} = 0, \quad \overrightarrow{\tau}(x,0) = \overrightarrow{\tau}^0(\gamma(x)). \quad (3.41)$$

The matrix $\tilde{A}$ is equal to $\{\frac{i}{2}\,\tau_i F_{,\tau_j}(\overrightarrow{\tau},t)\}_{1 \leq i,j \leq n}$. Note that $\mathrm{rank}\,\tilde{A} \leq 1$. By condition (H_2), either the function

$$\tilde{\lambda} = \mathrm{Tr}\,\tilde{A} = \sum_{1 \leq i \leq n} \frac{i}{2}\,\tau_i F_{,\tau_i}(\overrightarrow{\tau},0)$$

(the eigenvalue of $\tilde{A}$) is nonzero for all $x \in \mathbb{R}$, or $\tilde{A}(x) \equiv 0$. Hence (3.41) is hyperbolic for small enough t. In first case, the eigenvector of $\tilde{A}_{|t=0}$ for $\tilde{\lambda}(x)$ is $v_1 = (F_{,\tau_1}, F_{,\tau_2}, \ldots, F_{,\tau_n})$, and the kernel of $\tilde{A}_{|t=0}$ is spanned by $n-1$ vectors

$$v_2 = (-2\,\tau_2, \tau_1, 0, \ldots 0), \quad v_3 = (-3\,\tau_3, 0, \tau_1, 0, \ldots 0), \quad \ldots \quad v_n = (-n\,\tau_n, 0, \ldots 0, \tau_1).$$

(If $\tilde{\lambda}(x) = 0$ for some $x \in \mathbb{R}$, and $F_{,\tau_j}(\overrightarrow{\tau},0) \neq 0$ for some j, then $\tilde{A}$ is nilpotent and hence (3.41) is not hyperbolic).

By Theorem A, the initial value problem (3.41) has a unique solution on a domain $[-\delta,\delta] \times [0,\varepsilon')$ of the (x,t)-plane. Hence, there exists $t_q > 0$ such that the solution $\overrightarrow{\tau}(\cdot,t)$ to (3.28) exists and is unique for $t \in [0,t_q)$ on a neighborhood $U_q \subset M$ centered at q. By compactness of M, we conclude that there is an $\varepsilon > 0$ such that (3.29) admits a unique solution $\overrightarrow{\tau}(q,t)$ on M for $t \in [0,\varepsilon)$.

Using the properties of the Lie derivative of g_t along N, one may show that (3.29) is equivalent to

$$\mathscr{L}_{Z_t} g_t = 0 \quad \text{with} \ Z_t = \partial_t + \frac{1}{2} F(\overrightarrow{\tau}^{\,t}, t) N.$$

Denote by (Φ_t) the flow of $\frac{1}{2} F(\overrightarrow{\tau},t)N$. The solution metrics g_t can be determined by:

$$g_t(\Phi_t X, \Phi_t Y) = g_0(X,Y), \quad g_t(X,N) = 0, \quad g_t(N,N) = 1$$

for all X and Y tangent to $\mathscr{F}$. The solution exists and is unique as long as the solution to (3.28) does.   □

Remark 3.4. One may also apply the *method of characteristics* to solve (3.41) explicitly when $F = F(\overrightarrow{\tau})$. For $\tilde{\lambda} \neq 0$, the system has two characteristics:

$$\frac{d\tilde{x}}{dt} = \tilde{\lambda} + \frac{F}{2}, \qquad \frac{dx}{dt} = \frac{F}{2}$$

corresponding to two eigenvalues $\tilde{\lambda} + \frac{F}{2}$ (of multiplicity 1) and $\frac{F}{2}$ (of multiplicity $n-1$). First, let us observe that the function

$$u := v_1^T \cdot \overrightarrow{\tau} = \sum_{j \leq n} F_{,\tau_j} \tau_j$$

is constant along the first family of characteristics:

$$\frac{d}{dt} u = \partial_t u + \left(\tilde{\lambda} + \frac{F}{2} \right) N(u) = 0 \Leftrightarrow u = const \text{ along } \frac{d}{dt}\tilde{x} = \tilde{\lambda} + \frac{F}{2}. \tag{3.42}$$

Let us find a function that is constant along the second family of characteristics. For each $m > 1$, due to the form of v_m, calculate the sum of the first equation in (3.41) multiplied by $-m\,\tau_m$ with the mth equation multiplied by τ_1 (along the trajectories of $\frac{dx}{dt} = \frac{F}{2}$) and get

$$\tau_1 \partial_t \tau_m - m \tau_m \partial_t \tau_1 + \frac{F}{2} \left(\tau_1 N(\tau_m) - m \tau_m N(\tau_1) \right) = 0.$$

(The terms with $\sum_j F_{\tau_j} N(\tau_j)$ cancel). Using

$$\frac{d}{dt} \tau_m = \partial_t \tau_m + N(\tau_m) \frac{dx}{dt} = \partial_t \tau_m + \frac{F}{2} N(\tau_m),$$

we get (again, along the second family of characteristics)

$$\tau_1 \frac{d}{dt} \tau_m - m \tau_m \frac{d}{dt} \tau_1 = 0 \quad (m \geq 2). \tag{3.43}$$

The complete integral of (3.43) is

$$\tau_m = C_m(x)\, \tau_1^m \quad (m \geq 2), \tag{3.44}$$

where $C_m(x) = \tau_m(x,0)/\tau_1^m(x,0)$ are known functions. As the functions $\overrightarrow{\tau}(x,t)$ and $F(\overrightarrow{\tau})$ exist for $t \in [0,t_q)$, the EGF under consideration exists and is unique on $M \times [0,t_q)$.

Proof (of Corollary 3.2). Denote by $\overrightarrow{\tau}(x,t) = (\tau_1^t,\dots,\tau_n^t)$ the power sums of the principal curvatures of $\mathscr{F}$ at the point $\gamma(x)$ in time t. By Lemma 3.5, we have (3.2), which in our case reduces to (3.27). Using this system, we build the initial value problem in the (x,t)-plane

$$\partial_t \overrightarrow{\tau} + \tilde{A}(\overrightarrow{\tau},t)\partial_x \overrightarrow{\tau} = 0, \quad \overrightarrow{\tau}(x,0) = \overrightarrow{\tau}^{\,0}(\gamma(x)), \tag{3.45}$$

where $\tilde{A}$ is equal to $\{\frac{i}{2}\tau_{i-1}F_{,\tau_j}(\overrightarrow{\tau},t)\}_{1\le i,j\le n}$. As before, $\operatorname{rank}\tilde{A} \le 1$. Consider

$$\tilde{\lambda} = \operatorname{Tr}\tilde{A} = \sum_{1\le i\le n}\frac{i}{2}\tau_{i-1}F_{,\tau_i}(\overrightarrow{\tau},0)$$

(the eigenvalue of $\tilde{A}$). By (H_1), either $\tilde{\lambda}(x) \ne 0$ for all $x \in \mathbb{R}$ or $\tilde{A}(x) \equiv 0$. Hence, the system (3.45) is hyperbolic at $(x,0)$. As in the proof of Corollary 3.1 we conclude that there is an $\varepsilon > 0$ such that $\overrightarrow{\tau}(q,t)$ on M exists and is unique for $t \in [0,\varepsilon)$. Certainly, a unique solution to (3.30) is smooth and has the required form.   □

3.6 Global Existence of EGFs (Time Estimation)

Fix positive integers m and l, and put

$$J_{m,l} = \left\{\alpha \in \mathbb{Z}_+^n : \sum_j \alpha_j = m,\ \sum_j j\alpha_j = l\right\}.$$

For a vector $\alpha = (\alpha_1,\dots,\alpha_n) \in \mathbb{Z}_+^n$ put $\overrightarrow{\tau}^{\,\alpha} := \tau_1^{\alpha_1}\dots\tau_n^{\alpha_n}$.

Recall that a vector field on a manifold M is *complete* if any of its trajectory $\gamma : t \mapsto \gamma(t)$ can be extended to the whole range $\mathbb{R}$ of parameter t. If M carries a Riemannian structure g and a vector field X has bounded length then the completeness of (M,g) is sufficient for the completeness of X.

Proposition 3.3. *Let (M,g_0) be a Riemannian manifold with a codimension-one foliation $\mathscr{F}$ and a complete unit normal field N. Given $c_\alpha \in \mathbb{R}$ ($\alpha \in J_{m,l}$) and $m,l \in \mathbb{N}$, define functions $F_t = \sum_{\alpha\in J_{m,l}} c_\alpha(\overrightarrow{\tau}^t)^\alpha$ (where $\overrightarrow{\tau}^t$ is, as usual, the vector of power sums of principal curvatures of the leaves) and set*

$$T = \infty \ \text{if } N(F_0) \ge 0 \text{ on } M \text{ and } T = -\frac{2}{l+1}\Big/\inf_M N(F_0) \ \text{otherwise.}$$

Then the EGF $\partial_t g_t = F_t\,\hat{b}_1$, compare (3.29), has a unique smooth solution on M for $t \in [0,T)$ and does not possess one for $t \in [0,T]$.

Proof. Notice that

$$\sum_j (\vec{\tau}^{\,\alpha})_{,\tau_j}\tau_j = m\vec{\tau}^{\,\alpha}, \quad \sum_j j(\vec{\tau}^{\,\alpha})_{,\tau_j}\tau_j = l\,\vec{\tau}^{\,\alpha}.$$

If $F = \sum_\alpha c_\alpha \vec{\tau}^{\,\alpha}$ then $N(F) = \sum_j \sum_\alpha c_\alpha \vec{\tau}^{\,\alpha}_{,\tau_j} N(\tau_j)$ (the derivative of F along N).

Define $\tilde{F}(x,t) = F_t(\vec{\tau}(\gamma(x)))$ and $\tilde{F}_0 = \tilde{F}(\cdot,0)$. One has PDEs (3.41) in the (x,t)-plane. Characteristics of the first family, see (3.42), are lines and $\tilde{F} = const$ along them. To show this, observe that (by definition of $F_t(\vec{\tau})$)

$$u := \sum_j F_{t,\tau_j}(\vec{\tau})\,\tau_j = m\tilde{F}(x,t), \quad \tilde{\lambda} := \sum_j \frac{j}{2} F_{t,\tau_j}(\vec{\tau})\tau_j = \frac{l}{2}\tilde{F}(x,t).$$

Because $\tilde{F} = u/m$ is constant along the first family of characteristics in the (x,t)-plane, these characteristics (lines) are given by the equation

$$\frac{d}{dt}x = \frac{1}{2}(l+1)\tilde{F} \quad \Leftrightarrow \quad x = \xi + \frac{1}{2}(l+1)\tilde{F}_0(\xi)t.$$

Notice that

$$\tilde{F}' = \sum_{\alpha\in J_{m,l}} c_\alpha \sum_j (\vec{\tau}^{\,0})^{\alpha}_{,\tau_j} N(\tau_j^0).$$

If $\tilde{F}_0' > 0$ on γ, the solution $\tilde{F}$ exists for all $t \geq 0$ (see Example 3.2) and we set $t_q = \infty$. If $\tilde{F}_0'$ is negative somewhere on γ then $\tilde{F}$ exists (and is continuous) for $t \in [0,t_q)$ where

$$t_q = -2/[(l+1)\min_x \tilde{F}_0'(x)].$$

The second family of characteristics, $\frac{d}{dt}x = \frac{F}{2}$, also exists for $t \in [0,t_q)$. To show this, assume the contrary: there are $t_0 \in (0,t_p)$ and a trajectory $\gamma_1(t)$ of the second family of characteristics that cannot be continued for values $t \geq t_0$. Therefore, the inclination $\tilde{F}(\gamma_1(t),t)/2$ of γ_1 approaches to infinity when $t \to t_0$, a contradiction to continuity of $\tilde{F}$ on the strip $t \in [0,t_0]$ in the (x,t)-plane.   □

We shall apply EGFs to *umbilical* foliations (that is, those for which the Weingarten operator A is proportional to the identity at any point, among them *totally geodesic* foliations appear when $A = 0$) and to foliations on surfaces.

First we remark (see also Proposition 2.4) that EGFs preserve the umbilicity of $\mathscr{F}$.

Proposition 3.4. *Let (M,g_0) be a Riemannian manifold with umbilical foliation $\mathscr{F}$. If the EGF (3.1) of type (b) has a unique smooth solution g_t $(0 \leq t < \varepsilon)$ then $\mathscr{F}$ is umbilical for any g_t.*

Proof. As $\mathscr{F}$ is g_0-umbilical, we have $A_0 = \lambda_0\,\widehat{\mathrm{id}}$ for some function λ_0 on M. Let λ_t $(0 \le t < \tilde{\varepsilon})$ be a unique solution to the quasi-linear PDE

$$\partial_t\lambda_t + \frac{1}{2}N(\psi(\lambda_t,t)) = 0, \tag{3.46}$$

(see Theorem A with $n = 1$) where

$$\psi(\lambda,t) = \sum\nolimits_{j=0}^{n-1} f_j(n\lambda,n\lambda^2,\dots,n\lambda^n;t)\lambda^j$$

is the function of two variables. Consider the family of $\mathscr{F}$-truncated metrics $\tilde{g}_t$ $(0 \le t < \tilde{\varepsilon})$ defined along $\mathscr{F}$ by $\tilde{g}_t = g_0 \exp(\int_0^t \psi(\lambda_t,t)\,dt)$, hence $\partial_t \tilde{g}_t = \psi(\lambda_t,t)\tilde{g}_t$.

By Lemma 2.3, the Weingarten operator of $\mathscr{F}$ with respect to $\tilde{g}_t$ is conformal, $\tilde{A}_t = \mu_t\,\widehat{\mathrm{id}}$ for some function μ_t $(\mu_0 = \lambda_0)$. Hence $\mathscr{F}$ is $\tilde{g}_t$-umbilical.

By Lemma 2.9 with $S = s\hat{g}_t$ and $s = -\frac{1}{2}\psi$, we have

$$\partial_t\tilde{A}_t = -\frac{1}{2}N(\psi(\lambda_t,t)\widehat{\mathrm{id}}.$$

Taking the trace, we obtain the PDE

$$\partial_t\mu_t + \frac{1}{2}N(\psi(\lambda_t,t)) = 0.$$

Comparing with (3.46) we conclude that $\mu_t = \lambda_t$ for all t. Due to the definition of ψ, metrics $\tilde{g}_t$ also satisfy to (3.1). By uniqueness of the solution, we have $\tilde{g}_t = g_t$, which completes the proof. □

Proposition 3.5. *Let (M,g_0) be a Riemannian manifold, and $\mathscr{F}$ a codimension-one umbilical foliation on M with the normal curvature λ_0 and a complete unit normal field N. Then the EGF (3.1) with $h = \sum_{j=0}^{n-1} f_j(\overrightarrow{\tau})\hat{b}_j$ has a unique smooth solution g_t on M for $t \in [0,T)$, and does not possess one for $t \ge T$. Here*

$$T = \infty \text{ if } N(\psi'(\lambda_0)) \ge 0 \text{ on } M, \text{ and } T = -2/\inf_M N(\psi'(\lambda_0)) \text{ otherwise,}$$

where

$$\psi(\lambda) = \sum\nolimits_{j=0}^{n-1} f_j(n\lambda,\dots,n\lambda^n)\,\lambda^j. \tag{3.47}$$

Moreover, $\mathscr{F}$ is g_t-umbilical, $\hat{g}_t = \hat{g}_0 \exp(\int_0^t \psi(\lambda_t)dt)$, and λ_t is a unique smooth solution to the PDE

$$\partial_t\lambda_t + \frac{1}{2}N(\psi(\lambda_t)) = 0. \tag{3.48}$$

Proof. Theorem A (with $n = 1$) provides the short-time existence and uniqueness of the solution λ_t to (3.48) for $0 \leq t < T$. Furthermore, the EGF can be expressed as $\partial_t g_t = \psi(\lambda_t)\hat{g}_t$, and the solution $\hat{g}_t$ has the required form.

Consider the function $\tilde{\lambda}(x,t) = \lambda(\gamma(x),t)$ in the (x,t)-plane along the trajectory $\gamma(x)$, $\gamma(0) = q$, of N, and set $\tilde{\lambda}_0(x) = \lambda(\gamma(x),0)$. Equation (3.48) in this case has the form of a *conservation law*

$$\partial_t \tilde{\lambda} + \frac{1}{2}\partial_x(\psi(\tilde{\lambda})) = 0.$$

One may show the following.

If ψ', $\tilde{\lambda}_0 \in C^1(\mathbb{R})$ and if the functions $\tilde{\lambda}_0$ and ψ' are either nondecreasing or nonincreasing, the problem

$$\partial_t \tilde{\lambda} + \frac{1}{2}\partial_x(\psi(\tilde{\lambda})) = 0, \quad \tilde{\lambda}(x,0) = \tilde{\lambda}_0(x), \quad t \geq 0$$

has a unique smooth solution defined implicitly by the parametric equations,

$$\tilde{\lambda}(x,t) = \tilde{\lambda}_0(\xi), \quad x = \xi + \frac{1}{2}\psi'(\tilde{\lambda}_0(\xi))t.$$

If $\frac{d}{dx}\psi'(\tilde{\lambda}_0(x))$ is negative elsewhere along γ then $\tilde{\lambda}(x,t)$ exists for

$$t < t_q = -2 \Big/ \inf_{x \in \mathbb{R}} \frac{d}{dx}\psi'(\tilde{\lambda}_0(x)). \qquad\qquad \square$$

Notice that for $\psi = \lambda^2$, (3.48) reduces to Burgers' equation, see Example 3.2.

3.7 Variational Formulae for EGFs

In this section we shall use the EGF (3.1), whose f_j (at least one of them is not identically zero) are again of one of the two types:

$$(a)\ f_j = f_j(q)\quad (q \in M), \qquad (b)\ f_j = f_j(\overrightarrow{\tau}).$$

Under certain conditions the EGF admits a unique smooth solution g_t defined for some time interval $[0, \varepsilon)$, see Theorem 3.2 given earlier.

Recall that a self-adjoint $(1,1)$-tensor $h(A) = \sum_{j=0}^{n-1} f_j A^j$ is dual to $h(b)$ of (3.1). We shall apply the results of Sect. 2.3.2 with $f = \operatorname{Tr} h(A) = \sum_{j=0}^{n-1} f_j \tau_j$, to the functional

$$J_h : g \mapsto \int_M \operatorname{Tr} h(A)\, d\operatorname{vol}_g, \qquad g \in \mathcal{M}. \tag{3.49}$$

For umbilical foliations $\mathscr{F}$ (see Sect. 2.4.1) this reduces to the functional

$$I_\psi : g \mapsto \int_M \psi(\lambda)\,d\,\text{vol}_g, \qquad g \in \mathscr{M}. \tag{3.50}$$

3.7.1 The Normalized EGFs

A metric g on $(M, \mathscr{F})$ is a *fixed point* of the EGF (3.1) if it satisfies the condition

$$\sum_{j=0}^{n-1} f_j(\vec{\tau})A^j = 0.$$

For a generic setting of f_j's (when $f_0(0) = 0$), the *fixed points* of the EGF ($h(A) = 0$) are totally geodesic ($A_t \equiv 0$) foliations only. Several classes of foliations appear as fixed points of the flow $\partial_t g_t = f(\vec{\tau})\hat{g}_t$ for special choices of f, for example:

(a) Foliations of constant τ_i, when $f = \tau_i - c$ (minimal for $i = 1$ and $c = 0$).
(b) Umbilical foliations, when $f = n\,\tau_2 - \tau_1^2 = \sum_{i<j}(k_i - k_j)^2$ (see Example 2.6).
(c) Parabolic foliations, when $f = \sigma_n$.
(d) Totally geodesic foliations, when $f(0) = 0$, etc.

In order to extend the set of solutions (fixed points) we define the *normalized EGF* by:

$$\partial_t g_t = h(b_t) - (\rho_t/n)\,\hat{g}_t \quad \text{with} \quad \rho_t = J_h(g_t)\,/\,\text{vol}(M, g_t). \tag{3.51}$$

For the normalized EGF we have $\text{vol}(M, g_t) = const$, because, by (2.21),

$$\frac{d}{dt}\,\text{vol}(M, g_t) = \frac{1}{2}\int_M \text{Tr}\left(h(A) - (\rho_t/n)\,\widehat{\text{id}}\right)d\,\text{vol}_t = 0.$$

The EGF and its normalized companion provide some methods of evolving Riemannian metrics on foliated manifolds.

Fixed points of the normalized EGF satisfy $h(b_0) = (\rho/n)\hat{g}_0$, where $\rho = J_h(g)$. Among them there are umbilical ($b = \lambda g$) foliations with $\lambda = const$ on M.

Let g_t be a family of Riemannian metrics of finite volume on $(M, \mathscr{F})$. Metrics $\bar{g}_t = (\phi_t \hat{g}_t) \oplus g_t^\perp$ with $\phi_t = \text{vol}(M, g_t)^{-2/n}$ have unit volume: $\int_M d\overline{\text{vol}}_t = 1$.

Geometries (e.g., second fundamental forms b and $\bar{b}$) of metrics g_t and $\bar{g}_t$ are related by Lemma 2.3. Unnormalized and normalized EGF differ only by rescaling along the space $T\mathscr{F}$.

Proposition 3.6. *Let $(M, \mathscr{F})$ be a foliation, and g_t a solution (of finite volume) to the EGF (3.1) with $h(b) = \sum_{j=0}^{n-1} f_j(\vec{\tau})\hat{b}_j$. Then the metrics*

$$\bar{g}_t = (\phi_t\,\hat{g}_t) \oplus g^\perp, \quad \text{where} \quad \phi_t = \text{vol}(M, g_t)^{-2/n},$$

evolve according to the normalized EGF

$$\partial_t \bar{g}_t = h(\bar{b}_t) - (\rho_t/n)\,\hat{\bar{g}}_t, \quad \text{where } \rho_t = J_h(g_t)/\text{vol}(M, g_t). \tag{3.52}$$

Proof. By Lemma 2.3, $\bar{\tau}_j = \tau_j$ and $h(\bar{A}) = h(A)$ (respectively, for metrics $\bar{g}_t$ and g_t). Hence, $\text{Tr } h(\bar{A}) = \text{Tr } h(A)$. From (2.21) with $S = h(b_t)$ we get

$$\frac{d}{dt}\,\text{vol}(M, g_t) = \frac{d}{dt}\int_M d\,\text{vol}_t = \frac{1}{2}\int_M \text{Tr } h(A)\,d\,\text{vol}_t\,.$$

Thus, $\phi_t = \text{vol}(M, g_t)^{-2/n}$ is a smooth function. By Lemma 2.3, we have $h(\bar{b}_t) = \phi_t \cdot h(b_t)$. Therefore,

$$\partial_t \bar{g}_t = \phi_t \partial_t g_t + \phi'_t \hat{g}_t = h(\bar{b}_t) + \phi'_t/\phi_t\,\hat{\bar{g}}_t.$$

Using (2.21) and $d\overline{\text{vol}}_t = \phi_t^{n/2} d\,\text{vol}_t$, we obtain

$$\partial_t \overline{\text{vol}}_t = \partial_t(\phi_t^{\frac{n}{2}}\,\text{vol}_t) = \frac{n}{2}\,\phi_t^{\frac{n}{2}-1}\,\phi'_t\,\text{vol}_t + \frac{1}{2}\,\phi_t^{\frac{n}{2}}\,\text{Tr } h(A)\,\text{vol}_t = \frac{1}{2}\left(n\frac{\phi'_t}{\phi_t} + \text{Tr } h(A)\right)\overline{\text{vol}}_t\,.$$

Let ρ_t be the average of $\text{Tr } h(A)$, see (3.52). From the above we get

$$0 = 2\frac{d}{dt}\int_M d\,\overline{\text{vol}}_t = \int_M \left(n\frac{\phi'_t}{\phi_t} + \text{Tr } h(A)\right)d\,\overline{\text{vol}}_t = n\frac{\phi'_t}{\phi_t} + \rho_t.$$

This shows that $\rho_t/n = -\phi'_t/\phi_t$. Hence, $\bar{g}$ evolves according to (3.52). $\square$

Example 3.6. (a) For an umbilical foliation $\mathscr{F}$ (see Sect. 2.4.1) we have

$$h(A) = \psi(\lambda)\,\widehat{\text{id}}, \qquad \text{Tr } h(A) = n\,\psi(\lambda), \tag{3.53}$$

where λ is the normal curvature of $\mathscr{F}$, and $\psi(\lambda)$ is given in (3.48). The corresponding to EGF $\partial_t g_t = \psi(\lambda_t)\,\hat{g}_t$ the normalized companion, see (3.51), is

$$\partial_t g_t = (\psi - \rho_t)\hat{g}_t, \quad \text{where} \quad \rho_t = I_\psi(g_t)/\text{vol}(M, g_t). \tag{3.54}$$

For $\psi = \lambda$, due to $\int_M \lambda\,d\,\text{vol} = 0$, the EGF $\partial_t g_t = \lambda_t \hat{g}_t$ consist of metrics of the same volume.

(b) Obviously, the normalized EGF corresponding to $h(A) = A^k$ is defined by:

$$\partial_t g_t = \hat{b}_k^t - (\rho_t/n)\hat{g}_t \quad \text{with} \quad \rho_t = I_{\tau,k}(g_t)/\text{vol}(M, g_t). \tag{3.55}$$

(c) Similarly, the kth *normalized Newton transformation flow* is given by:

$$\partial_t g_t = T_k(b_t) - \frac{\rho_t}{n}\hat{g}_t \quad \text{with} \quad \rho_t = (n-k)I_{\sigma,k}(g_t)/\text{vol}(M, g_t). \tag{3.56}$$

3.7.2 First Derivatives of Functionals

Using Theorem 2.1 with $f = \operatorname{Tr} h(A)$ and Corollary 2.4, we obtain

Proposition 3.7. *If $g_t \in \mathcal{M}_1$ $(0 \le t < \varepsilon)$ is a solution to the normalized EGF (3.51), the first derivative of the functional (3.49) is given by:*

$$J'_h(g_t) = -\frac{1}{2} J_h^2(g_t) + \int_M \left(\frac{1}{2} (\operatorname{Tr} h(A))^2 - \langle B_h, \nabla'_N h(b) \rangle \right) d\operatorname{vol}_{g_t}, \qquad (3.57)$$

where $B_h = \sum_{i=1}^n \frac{i}{2} (\operatorname{Tr} h(A))_{,\tau_i} \hat{b}_{i-1}$. Moreover, if $g_t \in \mathcal{M}_1$ is a solution to (3.54) (hence $\mathcal{F}$ is umbilical) then the first derivative of the functional (3.50) is given by:

$$2 I'_\psi(g_t) = -n I_\psi^2(g_t) + \int_M \left(n \psi^2(\lambda_t) - \psi'(\lambda_t) N(\psi(\lambda_t)) \right) d\operatorname{vol}_{g_t}. \qquad (3.58)$$

Proof. From (2.22) and (2.35) with $S = h(b) - (\rho/n)\hat{g}$ and $f = \operatorname{Tr} h(A)$, we have

$$J'_h(g_t) = \int_M \left\langle \frac{1}{2} (\operatorname{Tr} h(b) - J_h(g_t)) \hat{g}_t - \mathscr{V}(B_h), h(b) - \frac{1}{n} J_h(g_t) \hat{g}_t \right\rangle d\operatorname{vol}_{g_t},$$

$$I'_\psi(g_t) = \frac{n}{2} \int_M \left(\psi(\lambda_t) - I_\psi(g_t) - \frac{1}{n} \mathscr{V}(\psi'(\lambda_t)) \right) (\psi(\lambda_t) - I_\psi(g_t)) d\operatorname{vol}_{g_t}.$$

The claim follows. $\qquad\square$

From Proposition 3.7 for $h(b) = \hat{b}_k$ and $h(b) = T_k(b)$, respectively, we obtain

Corollary 3.4. *If $g_t \in \mathcal{M}_1$ $(0 \le t < \varepsilon)$ is one of the normalized EGFs (3.55) – (3.56), the first derivative (3.57) reduces, respectively, to*

$$2 I'_{\tau,k}(g_t) = -I_{\tau,k}^2(g_t) + \int_M \left(\tau_k^2 - \frac{k^2}{2k-1} \tau_1 \tau_{2k-1} \right) d\operatorname{vol},$$

$$2 I'_{\sigma,k}(g_t) = -(n-k) I_{\sigma,k}^2(g_t) + (n-k) \int_M \sigma_k^2 d\operatorname{vol} - \int_M \langle T_{k-1}(b), \nabla'_N T_k(b) \rangle d\operatorname{vol}.$$

Proof. For (3.55) with $h(A) = A^k$ and $S = \hat{b}_k - (\rho/n)\hat{g}$, we have $f = \tau_k$ and $B_h = \frac{k}{2} \hat{b}_{k-1}$. Equality (3.57) reduces to

$$I'_{\tau,k}(g_t) = \frac{1}{2} \int_M \left\langle \left(\tau_k - I_{\tau,k}(g_t) \right) \hat{g}_t - k \mathscr{V}(\hat{b}_{k-1}), \hat{b}_k - \frac{1}{n} I_{\tau,k}(g_t) \hat{g}_t \right\rangle d\operatorname{vol}$$

$$= -\frac{1}{2} I_{\tau,k}^2(g_t) + \frac{1}{2} \int_M \tau_k^2 d\operatorname{vol} - \frac{k}{2} \int_M \operatorname{Tr}(A^{k-1} \nabla_N A^k) d\operatorname{vol}.$$

From the above equality, using (3.4), (3.34), and Lemma 2.6, the formula for $I'_{\tau,k}(g_t)$ follows.

For (3.56) with $h(A) = T_k(A)$ and $S = T_k(b) - (\rho/n)\hat{g}$, we have $f = (n-k)\tau_k$ and $B_h = \frac{1}{2}(n-k)T_{k-1}(b)$. Equality (3.57) (divided by $n-k$) reduces to

$$I'_{\sigma,k}(g_t) = -\frac{1}{2}(n-k)I^2_{\sigma,k}(g_t) + \int_M \left(\frac{1}{2}(n-k)\sigma_k^2 - \frac{1}{2}\langle T_{k-1}(b), \nabla_N^t T_k(b) \rangle \right) d\,\mathrm{vol}.$$

From the above the required formula for $I'_{\sigma,k}(g_t)$ follows.   □

Example 3.7. Let us recall the formula

$$\mathscr{B}_N = I_{\tau,2} + \int_M \|Z\|_g^2 d\,\mathrm{vol}$$

for the total bending of the unit normal N (Sect. 2.4.2) and the definition $h(A) = \sum_{j=0}^{n-1} f_j A^j$ for the EGF's evolution operator, see (3.1). Using (2.39), we obtain the following corollary.

Let $g_t \in \mathscr{M}_1$ be a solution to the normalized EGF (3.51), N – a unit normal to $\mathscr{F}$, and $Z = \nabla_N^t N$. By Theorem 2.2, the first derivative of the total bending is given by:

$$\mathscr{B}'_N(g_t) = \int_M \left\langle \frac{1}{2}(\|Z\|_t^2 + \tau_2 - \mathscr{B}_N(g_t))\hat{g}_t - Z^\flat \odot Z^\flat - \mathscr{V}(\hat{b}_1), h(b) - \frac{1}{n}\rho_t\hat{g}_t \right\rangle d\,\mathrm{vol}$$

$$= \int_M \left(\frac{1}{2}\operatorname{Tr} h(A)(\|Z\|_t^2 + \tau_2 - \mathscr{B}_N(g_t)) - h(b)(Z,Z) + \frac{1}{n}\rho_t\|Z\|_t^2 \right.$$

$$\left. + \sum_{j=0}^{n-1} f_j \left(\tau_1 \tau_{j+1} - \frac{1}{j+1}N(\tau_{j+1}) \right) \right) d\,\mathrm{vol}.$$

In Propositions 3.3 and 3.5 we provided sufficient conditions for existing and uniqueness of the EGF g_t for $t \in (0, \infty)$. Using this and Proposition 3.7, we obtain

Theorem 3.3. *(a) If $g_t \in \mathscr{M}_1 (t \geq 0)$ is a normalized EGF (3.51) on a closed manifold M and $h(b) = \tau_1^k \hat{b}_1 (k \in \mathbb{N})$ then g_t approach (in a weak sense) a metric making $\mathscr{F}$ a minimal foliation.*
(b) If $g_t \in \mathscr{M}_1 (t \geq 0)$ is a normalized EGF (3.53) with $\psi(\lambda) = \lambda^k (k \in \mathbb{N})$ and umbilical $\mathscr{F}$ then g_t approach (in a weak sense) a metric making $\mathscr{F}$ totally geodesic.

Proof. (a) We have $\operatorname{Tr} h(A) = \tau_1^{k+1}$ and $B_h = \frac{1}{2}(k+1)\tau_1^k \hat{g}$ (see Proposition 3.7). From (3.2) with $i = 1$ we conclude that

$$\partial_t \tau_1 = -\frac{1}{2}(k+1)\tau_1^k N(\tau_1).$$

By (3.57) and Lemma 2.5, the derivative of the functional $J_t = \int_M \tau_1^{k+1} d\,\mathrm{vol}_t$ satisfies the differential inequality

$$2J'_t = -\frac{1}{2}\left(\frac{k^2}{2k+1} \int_M \tau_1^{2k+2} d\,\mathrm{vol}_t + J_t^2 \right) \leq -\frac{1}{2}J_t^2.$$

If $\mathscr{F}$ is minimal w.r.t. some $g_{\bar{t}}$ (i.e., $\tau_1 \equiv 0$ at $t = \bar{t}$) then by Theorem A (with $n = 1$) $\mathscr{F}$ is minimal w.r.t. all g_t (i.e., $\tau_1 \equiv 0$ for all t) that completes the proof in this case.

Now, suppose that $\mathscr{F}$ is not minimal w.r.t. to any g_t, then $\int_M \tau_1^{2k+2} d\,\text{vol}_t > 0$, and certainly $J_t' < 0$, for all t. One may compare the above inequality $J_t' \le -\frac{1}{4}J_t^2$ with the Riccati ODE

$$y'(t) = -\frac{1}{4}y^2(t).$$

If $J_0 < 0$ then $J_t < 0$ for all $t \ge 0$, and $\lim_{t \to T} J_t = -\infty$ for some $T > 0$, – a contradiction. Similarly, we find that $J_t > 0$ for all $t \ge 0$. Comparing with the Riccati ODE, we see that the solution can be estimated as $J_t \le 4\big(t + 4/J_0\big)^{-1}$. As g_t are defined for $t \in [0, \infty)$, we have $J_t \to 0$ as $t \to \infty$.

(b) The proof is similar to (a). From (3.46) we conclude that

$$\partial_t \lambda = -\frac{1}{2}(k+1)\lambda^k N(\lambda).$$

By (3.58), and Lemma 2.5, the derivative of the functional $I_{\psi,t} := I_\psi(g_t)$ satisfies

$$2I_{\psi,t}' = -n\left(\frac{(k-1)^2}{2k-1}\int_M \lambda^{2k} d\,\text{vol}_t + I_{\psi,t}^2\right) \le -n I_{\psi,t}^2.$$

If $\mathscr{F}$ is totally geodesic w.r.t. some $g_{\bar{t}}$ (i.e., $\lambda \equiv 0$ at $t = \bar{t}$) then by Theorem A (with $n = 1$) $\mathscr{F}$ is totally geodesic w.r.t. all g_t, that completes the proof in this case.

Now, suppose that $\mathscr{F}$ is not totally geodesic w.r.t. to any g_t. Then $\int_M \lambda^{2k} d\,\text{vol}_t > 0$, and certainly $I_{\psi,t}' < 0$, for all t. One may compare the above inequality $I_{\psi,t}' \le -\frac{1}{4}I_{\psi,t}^2$ with the Riccati ODE $y'(t) = -\frac{n}{2}y^2(t)$. If $I_{\psi,0} < 0$ then $I_{\psi,t} < 0$ for all $t \ge 0$, and $\lim_{t \to T} I_{\psi,t} = -\infty$ for some $T > 0$ – a contradiction. Similarly we find that $I_{\psi,t} > 0$ $(t \ge 0)$. Comparing with the Riccati ODE, we see that the solution is estimated as

$$I_{\psi,t} \le \left(\frac{n}{2}t + I_{\psi,0}^{-1}\right)^{-1}.$$

As g_t are defined for $t \ge 0$, we have $I_{\psi,t} \to 0$ as $t \to \infty$.   □

3.8 Extrinsic Geometric Solitons

Special *soliton* solutions of geometric flow motivate the general analysis of the singularity formation. In this section we introduce soliton solutions to EGFs and study their geometry for umbilical foliations, foliations on surfaces, and in the case when the EGF is produced by the extrinsic Ricci curvature tensor.

Fig. 3.2 Horosphere
foliation

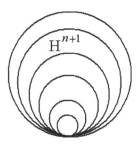

3.8.1 Introducing EGS

Let Diff(M) be the diffeomorphism group of M. Let us introduce the following notation:

– $\mathscr{D}(\mathscr{F})$ the subgroup of Diff(M) preserving $\mathscr{F}$.
– $\mathscr{D}(\mathscr{F},N)$ the subgroup of Diff($\mathscr{F}$) preserving both $\mathscr{F}$ and N.

Recall that the Fibration Theorem of D. Tischler (see, for example, [12]) states that the property of a closed manifold:

(a) *M admits a codimension one C^1-foliation invariant by a transverse flow*;
    is equivalent to each of the following conditions:
(b) *M fibers over the circle S^1.*
(c) *M supports a closed 1-form of a class C^1 without singularities.*
    Taking into account the theorem of R. Sacksteder, see again [12], in class C^2
    one obtains the following condition equivalent to any of (a), (b), or (c):
(d) *M admits a codimension one foliation without holonomy.*

Definition 3.4. We say that a solution $g_t = \hat{g}_t \oplus g_t^{\perp}$ to (3.1) is a *self-similar extrinsic geometric soliton* (EGS) on $(M,\mathscr{F},N)$ if there exist a smooth function $\sigma(t) > 0$ ($\sigma(0) = 1$), and a family of diffeomorphisms $\phi_t \in \mathscr{D}(\mathscr{F},N)$, $\phi_0 = \mathrm{id}_M$, such that

$$\hat{g}_t = \sigma(t)\,\phi_t^*\,\hat{g}_0. \tag{3.59}$$

A simple example of EGS appears on the hyperbolic space $\mathbb{H}^{n+1}$ with horosphere (horocycle when $n = 1$) foliation, see Sect. 3.9.4. On the Poincaré $(n+1)$-ball B the leaves of such Riemannian umbilical foliations are Euclidean n-spheres tangent to ∂B (Fig. 3.2). Trajectories orthogonal to the above foliations form foliations by geodesics.

Let $\mathbb{R}_+$ act on $\mathscr{M}(M,\mathscr{F},N)$ by scalings along $T\mathscr{F}$. By Remark 2.1, the Weingarten operator A and the principal curvatures of $\mathscr{F}$ are invariant under uniform scaling of the metric on $\mathscr{F}$. Therefore, EGFs may be regarded as dynamical systems on the quotient space $\mathscr{M}(M,\mathscr{F},N)/(\mathscr{D}(\mathscr{F},N) \times \mathbb{R}_+)$. Solutions to (3.1) of the form (3.59) correspond to *fixed points* of the above dynamical system.

Question 3.1. Given $(M, \mathscr{F}, N)$, N being a vector field transverse to $\mathscr{F}$, codim $\mathscr{F} = 1$, and f_j $(0 \le j < n)$ of class C^2, do there exist complete EGS metrics on M? If they exist, study their properties, classify them, etc.

The question is closely related to the basic problem in the theory of foliations mentioned in the Introduction.

We are looking for initial conditions that give rise to self-similar EGSs.

Vector fields represent diffeomorphisms infinitesimally. Let $\mathscr{X}(M)$ be the Lie algebra of all vector fields on M with the bracket operation. Let us introduce also the following notation:

- $\mathscr{X}(\mathscr{F})$, the set of vector fields on M preserving $\mathscr{F}$.
- $\mathscr{X}(\mathscr{F}, N)$, the set of vector fields on M preserving $\mathscr{F}$ and commuting with N.

By the Jacobi identity, $\mathscr{X}(\mathscr{F})$ and $\mathscr{X}(\mathscr{F}, N)$ are subalgebras of the Lie algebra $\mathscr{X}(M)$. Moreover, for any $X \in \mathscr{X}(\mathscr{F})$ (or $X \in \mathscr{X}(\mathscr{F}, N)$) there exists a family $\phi_t \in \mathscr{D}(\mathscr{F})$ (respectively, $\phi_t \in \mathscr{D}(\mathscr{F}, N)$) such that $X = d\phi_t/dt$ at $t = 0$. If $\phi_t \in \mathscr{D}(\mathscr{F}, N)$ then $\varphi_{t*} N = N \circ \varphi_t$. For $X \in \mathscr{X}(\mathscr{F}, N)$ generated by ϕ_t, the above yields $\mathscr{L}_X N = 0$.

Remark 3.5. (a) The following conditions are equivalent, [55]:

$$X \in \mathscr{X}(\mathscr{F}) \iff [X, T\mathscr{F}] \subset T\mathscr{F}.$$

By the above and the definition of $\mathscr{X}(\mathscr{F}, N)$, we conclude that

$$X \in \mathscr{X}(\mathscr{F}, N) \iff [X, T\mathscr{F}] \subset T\mathscr{F} \quad \text{and} \quad [X, N] = 0. \tag{3.60}$$

(b) By (a), a normal vector field $X = e^f N$ preserves $\mathscr{F}$ if and only if

$$\nabla^{\mathscr{F}} f = -\nabla_N N.$$

Here $\nabla^{\mathscr{F}} f$ is the $\mathscr{F}$-gradient of a function $f \in C^1(M)$. Indeed, using (3.60), we get

$$0 = g([X, Y], N) = -e^f g(\nabla_N N, Y) + Y(e^f), \qquad \forall Y \perp N.$$

In particular, $N \in \mathscr{X}(\mathscr{F}, N)$ if and only if $\mathscr{F}$ is a Riemannian foliation.

Recall that the *Lie derivative* of a $(0, p)$-tensor S with respect to a vector field X is given by

$$(\mathscr{L}_X S)(Y_1, \ldots, Y_p) = X(S(Y_1, \ldots, Y_p)) - \sum_{i=1}^{p} S(Y_1, \ldots, \mathscr{L}_X Y_i, \ldots, Y_p). \tag{3.61}$$

Using Proposition 3.1, we obtain

Proposition 3.8. *Let g_t be a self-similar EGS (see Definition 3.4). Then*

$$b_t = \sigma(t)\,\phi_t^* b_0, \qquad A_t = \phi_{t*}^{-1} A_0 \phi_{t*}, \qquad \overrightarrow{\tau}^t = \overrightarrow{\tau}^0 \circ \phi_t, \tag{3.62}$$

$$h(b_t) = \sigma(t)\,\phi_t^* h(b_0). \tag{3.63}$$

Differentiating (3.59) yields

$$h(b_t) = \sigma'(t)\,\phi_t^*\,\hat{g}_0 + \sigma(t)\phi_t^*(\mathscr{L}_{X(t)}\hat{g}_0), \tag{3.64}$$

where $X(t) \in \mathscr{X}(\mathscr{F},N)$ is a time-dependent vector field generated by the family ϕ_t. Since $h(b_t) = \sigma(t)\phi_t^* h(b_0)$, one may omit the pull-back in (3.64),

$$h(b_0) = \sigma'(t)/\sigma(t)\,\hat{g}_0 + \mathscr{L}_{X(t)}\hat{g}_0. \tag{3.65}$$

Motivated by the above, we have the following

Definition 3.5. A pair (g,X) consisting of a metric $g = \hat{g} \oplus g^{\perp}$ on $(M,\mathscr{F},N)$, and a complete vector field $X \in \mathscr{X}(\mathscr{F},N)$ satisfying for some $\varepsilon \in \mathbb{R}$ the condition

$$h(b) = \varepsilon\,\hat{g} + \mathscr{L}_X\hat{g}, \qquad \text{where} \quad h(b) = \sum_{0 \le j < n} f_j(\vec{\tau})\,\hat{b}_j \tag{3.66}$$

is called an *EGS structure*. We will say also that X is *the vector field along which the EGS flows*. If $X = \nabla F$ for some function $F \in C^1(M)$, we have a *gradient EGS structure*. In this case, $\frac{1}{2}\mathscr{L}_{\nabla F}\,\hat{g} = \widehat{\mathrm{Hess}}_g F$ (the $\mathscr{F}$-*truncated hessian*), and

$$h(b) = \varepsilon\,\hat{g} + 2\widehat{\mathrm{Hess}}_g F. \tag{3.67}$$

Remark 3.6. For a gradient EGS with $X \perp N$ one has $N(F) = 0$. For a gradient EGS with $X \| N$, the function F is constant along the leaves.

Equation (3.66) yields a rather strong condition on the EGS structure (g,X). For example, contracting (3.66) with g (tracing) and using the identity $\mathrm{Tr}\,\mathscr{L}_X\hat{g} = 2\,\mathrm{div}_{\mathscr{F}}X$ (see Lemma 3.7 in what follows) yields

$$\mathrm{Tr}\,_g h(b) = n\varepsilon + 2\,\mathrm{div}_{\mathscr{F}}X. \tag{3.68}$$

For a gradient EGS, (3.68) means $\mathrm{Tr}\,_g h(b) = n\varepsilon + 2\Delta_{\mathscr{F}}F$.

Proposition 3.9. *(a) Let (g,X) be an EGS structure on $(M,\mathscr{F})$ with a compact leaf L. Then*

$$n\varepsilon = \int_L \mathrm{Tr}\,_g h(b)\,\mathrm{d\,vol}_{g,L}/\mathrm{vol}(L,g). \tag{3.69}$$

(b) Let M be closed and either $X \perp N$ and $\nabla_N N = 0$ or $X \| N$ and $\tau_1 = 0$. Then

$$n\varepsilon = \int_M \mathrm{Tr}\,_g h(b)\,\mathrm{d\,vol}_g/\mathrm{vol}(M,g). \tag{3.70}$$

In particular, if (g,X) is an EGS structure on $(M,\mathscr{F})$ and $\mathscr{F}$ is a Riemannian foliation (i.e., $\nabla_N N = 0$), with minimal leaves (i.e., $\tau_1 = 0$), then (3.70) holds.

Proof. In case (a), integrating (3.68) over L and applying the Divergence Theorem we obtain (3.69). To prove (b), by Lemma 2.5 with $F = g(X,N)$, we obtain

$$\int_M N(g(X,N))\,d\,\mathrm{vol} = \int_M \tau_1 g(X,N)\,d\,\mathrm{vol}.$$

Next, we have $g(\nabla_N X, N) = Ng(X,N) - g(X, \nabla_N N)$, and by the above,

$$\int_M g(\nabla_N X, N)\,d\,\mathrm{vol}_g = \int_M \Big(\tau_1 g(X,N) - g(X, \nabla_N N)\Big)\,d\,\mathrm{vol}_g.$$

Therefore, in case (b), integrating (3.68) over M implies (3.70). □

Lemma 3.7. *For arbitrary vector fields $X \in \mathscr{X}(\mathscr{F}, N)$ and $Y_i \in \Gamma(T\mathscr{F})$, we have*

$$(\mathscr{L}_X \hat{g})(N, Y_i) = \hat{b}_1(X, Y_i) - \hat{g}(\nabla_{X^\perp} N, Y_i), \qquad (\mathscr{L}_X \hat{g})(N,N) = 0,$$
$$(\mathscr{L}_X \hat{g})(Y_1, Y_2) = \hat{g}(\nabla_{Y_1} X, Y_2) + \hat{g}(\nabla_{Y_2} X, Y_1).$$

Proof. Using the identity $\nabla g = 0$ and the definition of $\hat{g}$, we obtain

$$(\nabla_X \hat{g})(N,N) = (\nabla_X \hat{g})(Y_1, Y_2) = 0, \quad (\nabla_X \hat{g})(N, Y_i) = \hat{b}_1(X, Y_i) - \hat{g}(\nabla_{X^\perp} N, Y_i).$$

By the above and the definition (3.61), we have

$$(\mathscr{L}_X \hat{g})(Y_1, Y_2) = X(\hat{g}(Y_1, Y_2)) - \hat{g}([X, Y_1], Y_2) - \hat{g}(Y_1, [X, Y_2])$$
$$= \hat{g}(\nabla_{Y_1} X, Y_2) + \hat{g}(\nabla_{Y_2} X, Y_1),$$
$$(\mathscr{L}_X \hat{g})(N, Y_i) = -\hat{g}([X,N], Y_i) = \hat{b}_1(X, Y_i) - \hat{g}(\nabla_{X^\perp} N, Y_i).$$

Similarly, $(\mathscr{L}_X \hat{g})(N,N) = 0$. Notice that $(\mathscr{L}_N \hat{g})(N, Y_i) = -\hat{g}(\nabla_N N, Y_i)$. □

Proposition 3.10. *Equation (3.66) for EGS with $X = \mu N$ ($\mu : M \to \mathbb{R}_+$) reads:*

$$h(b) = \varepsilon \hat{g} - 2\mu\,\hat{b}_1.$$

For Riemannian foliations with the EGS structure $(g, \mu N)$ we evidently have $\mu \equiv 1$.

Proof. From (3.66) and Lemma 3.7 (for $X = \mu N$) we obtain

$$(\mathscr{L}_{\mu N} \hat{g})(Y_1, Y_2) = -2\mu\,\hat{b}_1(Y_1, Y_2), \qquad Y_i \perp N. \qquad □$$

Example 3.8. Let $h(b) = \hat{b}_1$ (i.e., $f_j = \delta_{j1}$). Any metric making $(M, \mathscr{F})$ a Riemannian foliation, is a steady EGS with $X = N$ (unit normal). Indeed, $N \in \mathscr{X}(\mathscr{F}, N)$ for a bundle-like metric g (see Lemma 3.7), and we have $h(b) = \frac{1}{2} \mathscr{L}_N \hat{g}$.

3.8.2   Canonical Form of EGS

Next, we observe that Definitions 3.4 and 3.5 are in fact equivalent.

Theorem 3.4. (a) If g_t is a self-similar EGS on $(M, \mathcal{F}, N)$ then there exists a vector field $X \in \mathcal{X}(\mathcal{F}, N)$ such that the metric g_0 satisfies (3.66). (b) Conversely, given vector field $X \in \mathcal{X}(\mathcal{F}, N)$ and a solution g_0 to (3.66), there is a function $\sigma(t) > 0$ and a family of diffeomorphisms $\phi_t \in \mathcal{D}(\mathcal{F}, N)$ such that a family of metrics g_t, defined by (3.59) on $(M, \mathcal{F})$, is a solution to (3.1).

Proof. (a) Recall that $\sigma(0) = 1$ and $\phi_0 = \mathrm{id}$. Let $X = \frac{\mathrm{d}}{\mathrm{d}t} \phi_t{}_{|t=0} \in \mathcal{X}(\mathcal{F}, N)$ be the vector field generated by diffeomorphisms ϕ_t. Then we have

$$h(b_0) = \partial_t g_t{}_{|t=0} = \partial_t \hat{g}_t{}_{|t=0} = \sigma'(0)\hat{g}_0 + \mathscr{L}_X \hat{g}_0.$$

This implies that g_0 and X satisfy (3.66) with $\varepsilon = \sigma'(0)$.

(b) Suppose that a pair (g_0, X) satisfies (3.66). Put $\sigma(t) = e^{\varepsilon t}$; hence $\sigma'(0) = \varepsilon$. Let $\psi_t \in \mathcal{D}(\mathcal{F}, N)$ with $\psi_0 = \mathrm{id}_M$ be a family of diffeomorphisms generated by X. A smooth family g_t of $\mathcal{F}$-truncated metrics on M, defined by $\hat{g}_t = \sigma(t)\psi_t^* \hat{g}_0$, is of the form (3.59). Moreover,

$$\partial_t g_t = \sigma'(t)\psi_t^*(\hat{g}_0) + \sigma(t)\psi_t^*(\mathscr{L}_X \hat{g}_0) = \frac{\sigma'(t)}{\varepsilon}\psi_t^*(\varepsilon \hat{g}_0 + \mathscr{L}_X \hat{g}_0) = \frac{\sigma'(t)}{\varepsilon}\psi_t^*(h(b_0)).$$

By Proposition 3.8, we have $f_j(\overrightarrow{\tau}^0 \circ \psi_t) = f_j(\overrightarrow{\tau}^0)$ for $t \geq 0$, and

$$\psi_t^*(h(b_0)) = \psi_t^* \left(\sum_{j=0}^{n-1} f_j(\overrightarrow{\tau}^0)\hat{b}_j^0 \right) = \sum_{j=0}^{n-1} f_j(\overrightarrow{\tau}^0 \circ \psi_t)\, \psi_t^* \hat{b}_j^0$$

$$= \sum_{j=0}^{n-1} f_j(\overrightarrow{\tau}^t)\sigma^{-1}(t)\,\hat{b}_j^t = \sigma^{-1}(t) h(b_t).$$

Hence $h(b_t) = \sigma(t)\,\psi_t^*(h(b_0))$, and we conclude that $\partial_t g_t = \frac{\sigma'(t)}{\sigma(t)\varepsilon} h(b_t) = h(b_t)$. $\square$

Remark 3.7. If (X, g) is an EGS structure then from Theorem 3.4 and Proposition 3.8 it follows that all the τ's are constant along X:

$$X(\tau_i) = 0, \qquad 1 \leq i \leq n. \tag{3.71}$$

Example 3.9. We will illustrate Theorem 3.4, case (b). Let (g_0, X) with $X = 0$ be an EGS structure on $(M, \mathcal{F})$. Then $h(b_0) = \varepsilon \hat{g}_0$ for some $\varepsilon \in \mathbb{R}$. The family of leafwise conformal metrics $g_t = (e^{\varepsilon t}\hat{g}_0) \oplus g_0^{\perp}$ obviously satisfies the PDE $\partial_t g_t = \varepsilon \hat{g}_t$. Using $\hat{b}_j^t = e^{\varepsilon t}\,\hat{b}_j^0$ and $\overrightarrow{\tau}^t = \overrightarrow{\tau}^0$ (see Remark 2.1), we obtain

$$h(b_t) = e^{\varepsilon t}h(b_0) = e^{\varepsilon t}\,\varepsilon \hat{g}_0 = \varepsilon \hat{g}_t.$$

Hence $\partial_t g_t = h(b_t)$, and g_t is a self-similar EGS.

Theorem 3.5 (Canonical form). *Let a self-similar EGS* (g_t) *be unique among soliton solutions to (3.1) with initial metric* g_0. *Then there is a 1-parameter family of diffeomorphisms* $\psi_t \in \mathcal{D}(\mathcal{F}, N)$ *and a constant* $\varepsilon \in \{-1, 0, 1\}$ *such that*

$$\hat{g}_t = (1 + \varepsilon t) \, \psi_t^* \hat{g}_0. \tag{3.72}$$

The cases $\varepsilon = -1, 0, 1$ in (3.72) correspond respectively to *shrinking, steady,* or *expanding* EGS.

Proof. Similar to that of [14, Proposition 1.3]. For the convenience of the reader, we prove the Theorem in the case $\sigma''(0) \neq 0$. From (3.65) it follows that:

$$h(b_0) = (\log \sigma)'(t) \, \hat{g}_0 + \mathcal{L}_{X(t)} \hat{g}_0, \tag{3.73}$$

where $X(t) \in \mathcal{X}(\mathcal{F}, N)$ is a family of vector fields such that $X(t) = d\phi_t/dt$. Differentiating (3.73) with respect to t gives

$$(\log \sigma)''(t) \, \hat{g}_0 + \mathcal{L}_{X'(t)} \hat{g}_0 = 0. \tag{3.74}$$

Let $Y_0 = -X'(0)/(\log \sigma)''(0)$. We then have $\mathcal{L}_{Y_0} \hat{g}_0 = \hat{g}_0$. Substituting this into (3.65), we have for all t

$$h(b_0) = \mathcal{L}_{(\log \sigma)'(t)Y_0 + X(t)} \, \hat{g}_0.$$

Put $X_0 = (\log \sigma)'(0)Y_0 + X(0)$. Then $h(b_0) = \mathcal{L}_{X_0} \hat{g}_0$. Let $\psi_t \in \mathcal{D}(\mathcal{F}, N)$ be a family of diffeomorphisms generated by X_0. We will check that

$$\tilde{g}_t = (\psi_t^* \hat{g}_0) \oplus (\psi_t^* g_0)^{\perp} \tag{3.75}$$

is the EGF with the same initial conditions g_0, and that it is a steady soliton (i.e., $\sigma(t) = 1$ for all t). Indeed, differentiating (3.75), we have by (3.63),

$$\partial_t \tilde{g}_t = \psi_t^* (\mathcal{L}_{X_0} \hat{g}_0) = \psi_t^* (h(b_0)) = h(\psi_t^* b_0) = h(\tilde{b}_t).$$

Thus $g_t = \tilde{g}_t$, by uniqueness assumption for EGS solutions to our flow with initial metric g_0. Replacing ϕ_t by ψ_t we get $\sigma(t) \equiv 1$ in (3.59). $\qquad \square$

3.8.3 Umbilical EGS

Let $\mathcal{F}$ be an umbilical foliation on (M, g) with the normal curvature λ. We have $A = \lambda \, \widehat{\mathrm{id}}$ and $\tau_j = n\lambda^j$. By Propositions 3.4 and 3.5, EGFs preserve the umbilicity of $\mathcal{F}$ and is given by:

$$\partial_t g_t = \psi(\lambda_t) \, \hat{g}_t, \tag{3.76}$$

where ψ is given by (3.47). In this case, λ_t obeys the quasilinear PDE (3.48), and the EGS structure equation (3.66) reduce to the PDE

$$\psi(\lambda) = \varepsilon + (2/n)\,\mathrm{div}_{\mathscr{F}}\tilde{X}. \tag{3.77}$$

If (g,X) is an EGS structure with an umbilical metric, by (3.71) we have

$$X(\lambda) = 0. \tag{3.78}$$

Certainly, any one-dimensional foliation is totally umbilical. The EGF on a surface (M^2,g), foliated by curves $\mathscr{F}$, is given by (3.76), where $\lambda = \tau_1$ is the geodesic curvature of the curves-leaves, and $\psi = f_0 \in C^2(\mathbb{R})$. EGS equation (3.66) on $(M^2,\mathscr{F})$ reduce to the PDE

$$\psi(\lambda) = \varepsilon + 2\,\mathrm{div}_{\mathscr{F}}X, \tag{3.79}$$

see (3.77) with $n = 1$.

Remark 3.8. Let $\mathscr{F}$ be an umbilical foliation with metric

$$g_t = (e^{2f_t}\hat{g}_0) \oplus g_0^{\perp}, \tag{3.80}$$

where $f_t : M \to \mathbb{R}$ $(f_0 = 0)$ are smooth functions. We claim that

$$2\,\partial_t f_t = \psi\big(\lambda_0 - N(f_t)\big), \tag{3.81}$$

where λ_0 is the normal curvature of g_0. Indeed, by Lemma 2.3, $A_t = A_0 - N(f_t)\,\widehat{\mathrm{id}}$, hence, $\lambda_t = \lambda_0 - N(f_t)$. By Lemma 2.3, we also have

$$b_t = e^{2f_t}(\lambda_0 - N(f_t))\,\hat{g}_0 = (\lambda_0 - N(f_t))\,\hat{g}_t.$$

Similarly, $b_t^j = (\lambda_0 - N(f_t))^j\,\hat{g}_t$. Differentiating (3.80) yields

$$\psi(\lambda_t)\hat{g}_t = h(b_t) = \partial_t g_t = (2\,\partial_t f_t)\,\hat{g}_t.$$

Hence, $2\,\partial_t f_t = \psi(\lambda_t)$ that gives us the nonlinear PDE (3.81).

One can solve this equation explicitly for f_t in the special case $f = c_1\lambda + c_2$ of the problem $\partial_t g_t = \psi(\lambda_t)\hat{g}_t$, when (3.81) becomes linear of the form

$$2\,\partial_t f_t + N(f_t) = c_1\lambda_t + c_2.$$

In general, the nonlinear PDE (3.81) is difficult to solve, and we apply EGF's approach: first we find λ_t from (3.48) then find g_t from (3.76).

Example 3.10. Let $\mathscr{F}$ be an umbilical foliation on (M,g_0) with $\lambda = \mathrm{const}$ (if M is closed then $\lambda = 0$ by the known IF (1.2), i.e., $\int_M \lambda f\,\mathrm{vol} = 0$), and X – an *infinitesimal homothety along leaves:* $\mathscr{L}_X \hat{g} = C\hat{g}$, where $C \in \mathbb{R}$. Using $h(b_0) = \psi(\lambda)\hat{g}_0$, see (3.76) and (3.5), we conclude that (g_0, X) is an EGS structure with $\varepsilon = \psi(\lambda) - C$. Taking $X = 0$, we obtain a self-similar EGS,

$$g_t = (e^{\psi(\lambda)t}\hat{g}_0) \oplus g_0^\perp \quad \text{with} \quad \phi_t = \mathrm{id}_M.$$

For the special case of a totally geodesic foliation (i.e., $\lambda \equiv 0$, if such g_0 exists on $(M, \mathscr{F})$), the pair (g_0, X) is an EGS structure with $\varepsilon = \psi(0) - C$. In particular, the totally geodesic metrics $g_t = (e^{\psi(0)t}\hat{g}_0) \oplus g_0^\perp$ provide a self-similar EGS with $\phi_t = \mathrm{id}_M$.

Conversely, let (g, X) be an EGS structure on $(M, \mathscr{F})$. If $\mathscr{F}$ is umbilical (with the normal curvature λ) then X is a *leaf-wise conformally Killing field*:

$$\mathscr{L}_X \hat{g} = (\psi(\lambda) - \varepsilon)\,\hat{g}.$$

If $\mathscr{F}$ is totally geodesic (hence $\psi = f_0(0)$), then X is an infinitesimal homothety along leaves, $\mathscr{L}_X \hat{g} = C\hat{g}$, with the factor $C = f_0(0) - \varepsilon$. In particular, X is a *leaf-wise Killing field*, $\mathscr{L}_X \hat{g} = 0$, when $\varepsilon = f_0(0)$. This happens, for example, when M is a surface of revolution in $M^{n+1}(c)$ foliated by parallels, see Example 3.16.

Example 3.11. Consider biregular foliated coordinates (x_0, x_1) on a surface M^2 (see [15, Sect. 5.1]). Because the coordinate vectors ∂_0, ∂_1 are directed along N and $\mathscr{F}$, respectively, the metric has the form $g = g_{00}\,dx_0^2 + g_{11}\,dx_1^2$. Recall that $h(b) = \psi(\lambda)g_{11}$. By Lemma 2.5 with $n = 1$, we have

$$\lambda = -\frac{1}{2\sqrt{g_{00}}}(\log g_{11})_{,0}.$$

Let $X = X^0 \partial_0 + X^1 \partial_1 \in \mathscr{X}(\mathscr{F}, N)$. Using $g_{01} = 0$, we obtain

$$\mathrm{div}_{\mathscr{F}} X = g(\nabla_{\partial_1} X, \partial_1) = (\partial_1(X^1) + X^0 \Gamma_{01}^1 + X^1 \Gamma_{11}^1)g_{11},$$

where $\Gamma_{01}^1 = \frac{1}{2}(\log g_{11})_{,0}$ and $\Gamma_{11}^1 = \frac{1}{2}(\log g_{11})_{,1}$. Hence, (3.79) has the form

$$\psi(\lambda) - \varepsilon = 2\partial_1(X^1)g_{11} + X^0 g_{11,0} + X^1 g_{11,1}.$$

From the condition $[X, \partial_1] \perp \partial_0$, see (3.60), and

$$[X, \partial_1] = -\partial_1(X^0)\partial_0 - \partial_1(X^1)\partial_1$$

we obtain $\partial_1(X^0) = 0$. Next, from the condition $[X, N] = 0$, see (3.60) again, and

$$\left[X, g_{00}^{-\frac{1}{2}}\partial_0\right] = \left(X(g_{00}^{-\frac{1}{2}}) - g_{00}^{-\frac{1}{2}}\partial_0(X^0)\right)\partial_0 + g_{00}^{-\frac{1}{2}}\partial_0(X^1)\partial_1$$

we obtain

$$\partial_0(X^1) = 0, \qquad \partial_0(X^0) = -\frac{1}{2}X(\log g_{00}).$$

Definition 3.6 (see [31]). Denote the torus $\mathbb{R}^{n+1}/\mathbb{Z}^{n+1}$ by T^{n+1}. For $v \in \mathbb{R}^{n+1}$, let $R_v^t(x) := x + tv$ be the flow on T^{n+1} induced by a "constant" vector field X_v. We say $v \in \mathbb{R}^{n+1}$ is *Diophantine* if there is $s > 0$ such that

$$\inf\{|\langle u, v\rangle| \cdot \|u\|^s > 0 : u \in \mathbb{Z}^{n+1} \setminus \{0\}\},$$

where $\langle\,,\,\rangle$ and $\|\cdot\|$ are the Euclidean inner product and the norm in $\mathbb{R}^{n+1}$. When v is Diophantine, we call R_v a *Diophantine linear flow*.

Theorem 3.6. *Let $\mathscr{F}$ be an umbilical foliation (with a unit normal N) on a torus (T^{n+1}, g), $n > 0$. Suppose that X is a smooth unit vector field on $T\mathscr{F}$ with the properties*

$$\text{(i) } \nabla_X X \perp T\mathscr{F}, \qquad \text{(ii) } R(X,Y)Y \in T\mathscr{F} \qquad (Y \in T\mathscr{F}). \qquad (3.82)$$

If the X-flow is conjugate (by a homeomorphism) to a Diophantine linear flow R_v then for any function ψ of a class C^2 in either (3.76) or (3.47), there exists $f : \mathrm{T}^{n+1} \to \mathbb{R}$ such that (g, fX) is an EGS structure, that is, (3.77) holds with X replaced by fX.

Proof. For an umbilical foliation with the normal curvature λ, the Weingarten operator is conformal: $A = \lambda\,\hat{\mathrm{id}}$. By $(3.82)_{ii}$ and the *Codazzi equation* (see [42])

$$(\nabla_X A)Y - (\nabla_Y A)X = R(X,Y)N^\top,$$

we have $\lambda = const$ along X-curves. By $(3.82)_i$, the X-curves are $\mathscr{F}$-geodesics.

Let $(g, \tilde{X})$ be an EGS structure with $\tilde{X} = fX$. From (3.77) and the known identity

$$\mathrm{div}_{\mathscr{F}}(fX) = f\,\mathrm{div}_{\mathscr{F}} X + X(f)$$

it follows that $\mathrm{div}_{\mathscr{F}} \tilde{X} = X(f)$. We are looking for a solution of PDE, see (3.79),

$$\psi(\lambda) - \varepsilon = (2/n)X(f).$$

As X-flow is conjugate to a Diophantine linear flow, by the Kolmogorov Theorem (see [31]) the above PDE has a solution $(f, \varepsilon) \in C^\infty(\mathrm{T}^{n+1}) \times \mathbb{R}$. $\qquad\square$

From Theorem 3.6 follows:

Corollary 3.5. *Let a unit vector field X on a torus (T^2, g) define a foliation $\mathscr{F}$ by curves of constant geodesic curvature λ. If X-flow is conjugated to a Diophantine liner flow R_v then for any function ψ of a class C^2 in (3.76), (3.47), there exists $f : \mathrm{T}^2 \to \mathbb{R}$ such that (g, fX) is an EGS structure, see (3.79).*

Notice that, if $\psi \in C^2(\mathbb{R})$ then the following function is differentiable of class C^1:

$$\mu = \begin{cases} -\frac{n}{2}(\psi(\lambda) - \psi(0))/\lambda, & \lambda \neq 0, \\ -\frac{n}{2}\psi'(0), & \lambda = 0. \end{cases} \tag{3.83}$$

Theorem 3.7. *Let $\mathscr{F}$ be an umbilical foliation on (M, g) with normal curvature λ, and the function $\psi \in C^2(\mathbb{R})$ given in (3.47) satisfies $\psi' \neq 0$. Then the following properties are equivalent:*

(1) The normal curvature of $\mathscr{F}$ satisfies $N(\lambda) = 0$.
(2) $(g, \mu N)$, for some function μ, is an EGS structure, compare (3.77), indeed, one may take μ as in (3.83) and $\varepsilon = \psi(0)$.

Proof. $(1) \Rightarrow (2)$: The EGS equations (for an umbilical metric g and the vector field $X = \mu N$) are, see Proposition 3.10 and (3.78),

$$\psi(\lambda) - \varepsilon = -(2/n)\mu\lambda, \qquad X(\lambda) = 0. \tag{3.84}$$

For $\varepsilon = \psi(0)$ and μ given in (3.83), the above (3.84) are satisfied. Hence, by Definition 3.5, the pair $(g, \mu N)$ satisfies (3.77).

$(2) \Rightarrow (1)$: Using Definition 3.5, (3.78) and $\psi' \neq 0$, we have the equality $\mu N(\lambda) = 0$ with μ given in (3.83). Consider an open set

$$\Omega = \text{int}\{q \in M : \mu(q) = 0\}.$$

Certainly, $N(\lambda) = 0$ on $M \setminus \Omega$. By (3.83), we have $\psi(\lambda) = \psi(0)$. Thus $N(\psi(\lambda)) = \psi'(\lambda)N(\lambda) = 0$ on Ω. Since $\psi' \neq 0$, we have $N(\lambda) = 0$ on Ω. From the above we conclude that $N(\lambda) = 0$ on M. $\qquad\square$

From Theorem 3.7 it follows directly

Corollary 3.6. *Let $\mathscr{F}$ be a foliation (by curves) on a surface (M^2, g), $\psi \in C^2(\mathbb{R})$ be as in (3.47) and satisfies $\psi' \neq 0$. Then the following properties are equivalent:*

(1) The geodesic curvature λ of $\mathscr{F}$ satisfies $N(\lambda) = 0$.
(2) $(g, \mu N)$, for some function μ, is an EGS structure, compare (3.79), indeed, one may take μ as in (3.83) with $n = 1$ and $\varepsilon = \psi(0)$.

Example 3.12 (Non-Riemannian EGS on double-twisted products). Let $M = M_1 \times M_2$ be the product (with the metric $\tilde{g} = g_1 \oplus g_2$ and Levi-Civita connection $\tilde{\nabla}$) of a closed Riemannian manifold (M_1, g_1) and a circle $M_2 = S^1$ with the canonical metric g_2. Let $f_i : M \to \mathbb{R}$ $(i = 1, 2)$ be positive differentiable functions, $\pi_i : M \to M_i$ the canonical projections, $\pi_{i*} : TM \to \ker \pi_{3-i}$ the vector bundle projections. The *metric of a double-twisted product $M_1 \times_{(f_1, f_2)} M_2$ is given by:*

$$g(X, Y) = f_1^2 g_1(\pi_{1*}X, \pi_{1*}Y) + f_2^2 g_2(\pi_{2*}X, \pi_{2*}Y), \quad X, Y \in TM,$$

i.e., $g = (f_1^2 g_1) \oplus (f_2^2 g_2)$. The Levi-Civita connection ∇ of g obeys the relation [38]

$$\nabla_X Y = \tilde{\nabla}_X Y + \sum_{i=1,2} \left(g(\pi_{i*}X, \pi_{i*}Y) U_i - g(X, U_i) \pi_{i*}Y - g(Y, U_i) \pi_{i*}X \right), \quad (3.85)$$

where $U_i = -\nabla(\log f_i)$. Both foliations $M_1 \times \{q_2\}$ and $\{q_1\} \times M_2$ are umbilical with *mean curvature vectors* $H_1 = \pi_{2*}U_1$ and $H_2 = \pi_{1*}U_2$, respectively. This property characterizes the double-twisted product, see [38].

By Proposition 3.4, the EGFs preserve the above double-twisted product structure. The mean curvature of the foliation $\mathscr{F} := M_1 \times \{q_2\}$ is constant along the curves $\{q_1\} \times M_2$ (i.e., N-curves) if and only if $\pi_{2*}U_1$ is a function of M_1. In this case, by Theorem 3.7, $\mathscr{F}$ admits an EGS structure with $X \| N$. Remark that for $f_1 = e^\varphi$ and $f_2 = 1$, by (3.85), we find $A_N = -N(\varphi)\widehat{\mathrm{id}}$, as in Lemma 2.3.

Theorem 3.8. *Let (g, X) be an EGS structure on a closed surface M^2 foliated by curves $\mathscr{F}$ of the geodesic curvature λ (Torus T^2 is the only possibility here if we assume orientability!) Let $\psi \in C^2(\mathbb{R})$, see (3.47), satisfies $\psi' \neq 0$. If $X \| N$ then X-curves are closed and define a fibration $\pi : M^2 \to S^1$, and $\mathscr{F}$ is the suspension of a diffeomorphism $f : S^1 \to S^1$. Moreover, if $\psi(\lambda) = -2\lambda + c$ then (g, X) is the EGS structure (with $\varepsilon = c$) for any metric $g \in \mathscr{M}$ satisfying $N(\lambda) = 0$, otherwise $\lambda = 0$ (i.e., $\mathscr{F}$ is a geodesic foliation).*

Proof. Assume the contrary. Then the foliation $\mathscr{F}_N$ has a limit cycle. As M is compact, there is a domain $\Omega \subset M^2$ bounded by closed N-curves (which are limit cycles). By (3.78), $\lambda = const$ along N-curves. As there are limiting leaves, $\lambda = const$ on Ω. From the relation $\operatorname{div} N = -\lambda$ and the Divergence Theorem

$$\int_\Omega \operatorname{div} N \, \mathrm{d} \operatorname{vol} = \int_{\partial \Omega} \langle N, \nu \rangle \, \mathrm{d}\omega,$$

where ν is the outer normal to the boundary $\partial \Omega$ (hence $\nu \perp N$), we conclude that $\lambda = 0$ on Ω, hence $\mathscr{F}_N$ is a Riemannian foliation – a contradiction.

By the classification theorem for foliations on closed surfaces, see [22], all the X-curves are closed and define a fibration $\pi : M^2 \to S^1$. By (3.84) with $\mu = 1$ we have the following. If $\psi(\lambda) \neq -2\lambda + c$ then $\lambda = const$ on M. Hence, using the integral formula $\int_M \lambda \, \mathrm{d} \operatorname{vol} = 0$, we conclude that $\lambda = 0$. In this case, by Lemma 2.3, any $\mathscr{F}$-geodesic metric on $(M, \mathscr{F}, N)$ has the form $\bar{g} = (\pi^{-1} \circ \sigma \hat{g}) \oplus g^\perp$, where $\sigma : S^1 \to \mathbb{R}$ is a smooth function. $\qquad \square$

3.9 Applications and Examples

3.9.1 Extrinsic Ricci Flow

The *extrinsic Riemannian curvature tensor* $\mathrm{Rm}^{\mathrm{ex}}$ of $\mathscr{F}$ is, roughly speaking, the difference of the curvature tensors of M and of the leaves. More precisely, by the Gauss formula, see (1.5), we have

$$\mathrm{Rm}^{\mathrm{ex}}(X,Y)V = g(AY,V)AX - g(AX,V)AY.$$

Here, we study (in small dimensions $n > 1$) the *extrinsic Ricci flow*

$$\partial_t g_t = -2\,\mathrm{Ric}_t^{\mathrm{ex}}, \tag{3.86}$$

where the *extrinsic Ricci* tensor is given by:

$$\mathrm{Ric}^{\mathrm{ex}}(X,Y) = \mathrm{Tr}\,\mathrm{Rm}^{\mathrm{ex}}(\cdot,X)Y, \quad X,Y \in T\mathscr{F}.$$

Hence,

$$\mathrm{Ric}^{\mathrm{ex}} = \tau_1 \hat{b}_1 - \hat{b}_2. \tag{3.87}$$

Therefore, $-2\,\mathrm{Ric}^{\mathrm{ex}}$ relates to $h(b)$ of (3.1) with $f_1 = -2\tau_1$, $f_2 = 2$ (others $f_j = 0$). For $n = 2$, we have $\mathrm{Ric}^{\mathrm{ex}} = \sigma_2\,\hat{g}$.

Example 3.13. By Lemma 3.5, for the extrinsic Ricci flow (3.86) we have

$$\partial_t \tau_i = i\,\tau_i N(\tau_1) + \tau_1 N(\tau_i) - \frac{2i}{i+1} N(\tau_{i+1}), \quad i > 0. \tag{3.88}$$

Putting $\tau_i = \sum_{j=1}^{n}(k_j)^i$ in (3.88), we obtain PDEs for the principal curvatures

$$\partial_t k_i = N(k_i(\tau_1 - k_i)), \quad i = 1,\dots,n \tag{3.89}$$

which for an umbilical $\mathscr{F}$ (i.e., $k_i = \lambda$) reduce to the Burgers' type PDE

$$\partial_t \lambda = 2\,(n-1)\,\lambda\,N(\lambda).$$

For $n = 2$, (3.89) takes the form of the system for k_1 and k_2 with equal RHS's,

$$\partial_t k_i = N(k_1 k_2), \quad i = 1,2,$$

hence $\partial_t(k_2 - k_1) = 0$ and the difference $k_2 - k_1$ does not depend on t. We may say that in this case the pointwise "distance from umbilicity" is constant in time.

Corollary 3.7. *There exists a unique solution g_t, $t \in [0, \varepsilon)$ (for some $\varepsilon > 0$), to the extrinsic Ricci flow (3.86) in both of the following cases:*

(i) $n = 2$, and $\tau_1 \neq 0$; in this case, $\hat{g}_t = \hat{g}_0 \exp\left(\int_0^t \sigma_2\,dt\right)$;
(ii) $n = 3$, and $|\sigma_1|^3 > 27|\sigma_3| > 0$.

Notice that if a (positive or negative) definite operator A is not proportional to $\widehat{\mathrm{id}}$ then the inequalities of Corollary 3.7 (ii) are satisfied.

Proof. It is sufficient to show that in conditions of our Corollary, (3.88) is hyperbolic in the t-direction.

Let $n = 2$. By equality $\tau_3 = \frac{3}{2}\tau_1\tau_2 - \frac{1}{2}\tau_1^3$, the matrix of 2-truncated system (3.88) is $C_2 = \begin{pmatrix} -2\tau_1 & 1 \\ -2\tau_1^2 & \tau_1 \end{pmatrix}$. One can see directly, or applying condition (H_1) to $\mathrm{Ric}^{\mathrm{ex}} = \sigma_2 \hat{g}$, that C_2 is strictly hyperbolic if $\tau_1 \neq 0$. (If $\tau_1 \neq 0$, C_2 has real eigenvalues $\lambda_1 = 0$, $\lambda_2 = -\tau_1$, and the left eigenvectors $v_1 = (-\tau_1, 1)$ and $v_2 = (-2\tau_1, 1)$. If $\tau_1 \equiv 0$, i.e., $\mathscr{F}$ is minimal, the matrix C_2 is nilpotent, hence it is not hyperbolic). By Corollary 3.2, $\hat{g}_t = \hat{g}_0 \exp\left(\int_0^t \sigma_2\, dt\right)$, where σ_2 exists for $0 \leq t < T$.

Remark that $\tau_1^2 - 2\tau_2 = -(k_1 - k_2)^2 = const$ along the first family of characteristics $\frac{dx}{dt} = 0$, hence the function $k_1 - k_2$ does not depend on t. Along the second family of characteristics $\frac{dx}{dt} = -\tau_1$ we have $k_1 k_2 = (\tau_1^2 - \tau_2)/2 = const$.

For $n = 3$, the matrix of 3-truncated system (3.88) is

$$C_3 = \begin{pmatrix} -2\tau_1 & 1 & 0 \\ -2\tau_2 & -\tau_1 & \frac{4}{3} \\ \tau_1^3 - \tau_3 - 3\tau_1\tau_2 & \frac{3}{2}(\tau_2 - \tau_1^2) & \tau_1 \end{pmatrix}.$$

Replacing τ-s by σ-s, see Remark 1.1, we obtain the characteristic polynomial $P_3 = \lambda^3 + 2\sigma_1\lambda^2 + \sigma_1^2\lambda + 4\sigma_3$. Substituting $\lambda = y - \frac{2}{3}\sigma_1$ into P_3 gives

$$P_3 = y^3 + py + q, \quad \text{where} \quad p = -(1/3)\sigma_1^2 \text{ and } q = 4\sigma_3 - (2/27)\sigma_1^3.$$

Depending on the sign of the discriminant $D = (q/2)^2 + (p/3)^3$, we have

$$D \begin{cases} > 0, & \text{one real and two complex roots,} \\ < 0, & \text{three different real roots,} \\ = 0, & \text{one real root with multiplicity three in the case } p = q = 0, \\ & \text{or a single and a double real roots when } (\frac{1}{3}p)^3 = -(\frac{1}{2}q)^2 \neq 0. \end{cases}$$

In our case, $D = \frac{4}{27}\sigma_3(27\sigma_3 - \sigma_1^3)$. The condition "three different real roots" is $D < 0 \Leftrightarrow |\sigma_1|^3 > 27|\sigma_3| > 0$. $\qquad\square$

Corollary 3.8. *Let (M, g_0) be a Riemannian manifold with a codimension-one umbilical foliation $\mathscr{F}$ of the normal curvature λ_0 and a complete unit normal field N. Set at $t = 0$*

$$T = \infty \text{ if } N(\lambda_0) \leq 0 \text{ on } M, \text{ and } T = 1/[4(n-1)\sup_M N(\lambda_0)] \text{ otherwise.}$$

Then the extrinsic Ricci flow (3.86) has a unique smooth solution g_t on M for $t \in [0, T)$, and does not possess one for $t \geq T$.

Proof. The function $\tilde{\lambda}_t(x) = \lambda(\gamma(x), t)$ along the trajectory $\gamma(x)$ ($\gamma(0) = q$), of N satisfies (3.48) with $\psi(\lambda) = 4(1-n)\lambda^2$ and initial value $\tilde{\lambda}_0(x) = \lambda(\gamma(x), 0)$. The statement follows from Proposition 3.5. $\qquad\square$

3.9.2 *Extrinsic Ricci Solitons*

Foliations satisfying $\mathrm{Ric}^{\mathrm{ex}} = 0$ are fixed points of extrinsic Ricci flow. That is, they have the property $A(A - \tau_1\,\widehat{\mathrm{id}}) = 0$, or equivalently,

$$k_j(k_j - \tau_1) = 0 \quad (1 \le j \le n)$$

for the eigenvalues k_j of A. Because $\tau_1 = \sum_j k_j$, from above it follows $k_j = 0$ for all j. Hence, all extrinsic Ricci flat foliations are totally geodesic.

In order to extend the set of such solutions we shall apply the normalized EGFs, see (3.51).

Definition 3.7. We call $g_t \in \mathscr{M}$ on a closed manifold M a *normalized extrinsic Ricci flow* if

$$\partial_t g_t = -2\,\mathrm{Ric}^{\mathrm{ex}}_t + (\rho^{\mathrm{ex}}_t/n)\,\hat{g}_t, \quad \rho^{\mathrm{ex}}_t = -4\int_M \sigma_2\,\mathrm{d\,vol}_t \Big/ \mathrm{vol}(M, g_t). \tag{3.90}$$

Following Definition 3.5, an *extrinsic Ricci soliton structure* is a pair (g, X) of a metric g on $(M, \mathscr{F})$, and a complete field $X \in \mathscr{X}(\mathscr{F}, N)$ satisfying for some $\varepsilon \in \mathbb{R}$

$$-2\,\mathrm{Ric}^{\mathrm{ex}} = \varepsilon\,\hat{g} + \mathscr{L}_X\hat{g}.$$

Remark 3.9. To explain ρ^{ex}_t in (3.90), we find the *extrinsic scalar curvature*:

$$\mathrm{Tr}\,\mathrm{Ric}^{\mathrm{ex}} = \mathrm{Tr}\,(\tau_1 A - A^2) = \tau_1^2 - \tau_2 = 2\,\sigma_2.$$

Substituting this into (3.51) instead of $\int_M \mathrm{Tr}\,h(A)\,\mathrm{d\,vol}$, we obtain ρ^{ex}_t of (3.90). Notice that by the known integral formula (1.3), we have

$$\int_M \mathrm{Tr}\,\mathrm{Ric}^{\mathrm{ex}}\,\mathrm{d\,vol} = \int_M \mathrm{Ric}(N, N)\,\mathrm{d\,vol}.$$

Remark 3.10. A foliation (M, g) will be called *CPC (constant principal curvatures)* if the principal curvatures of leaves are constant.

(a) From (3.2) it follows that all the (either normalized of not) EGFs preserve CPC property of foliations. Indeed, let such a flow on $(M, \mathscr{F})$ starts from a CPC metric. From $N(\tau_i) = 0$, $N(f_j) = 0$ $(j < n)$, and (3.2) we conclude that τ's do not depend on t.

(b) Let (G, g) be a compact Lie group with a left invariant metric g. Suppose that the corresponding Lie algebra has a codimension one subspace V such that $[V, V] \subset V$. Then V determines a CPC foliation on (G, g).

Theorem 3.9. *Let (g, X) be an extrinsic Ricci soliton structure on $(M^n, \mathscr{F})$ $(n > 2)$, and X a leaf-wise conformal Killing field (i.e., $\mathscr{L}_X\hat{g} = \mu\,\hat{g}$).*

(i) Then, there are at most two distinct principal curvatures at any point $q \in M$.
(ii) Moreover, if μ is constant along the leaves then $\mathscr{F}$ is CPC foliation.

Proof. (i) As

$$(\mathrm{Ric}^{\mathrm{ex}})^{\sharp} = -A(A - \tau_1 \,\widehat{\mathrm{id}})$$

and $\mathscr{L}_X \hat{g} = \mu \, \hat{g}$, we obtain the equality $A(A - \tau_1 \,\widehat{\mathrm{id}}) = r \,\widehat{\mathrm{id}}$ with $r = \frac{1}{2}(\varepsilon + \mu)$, that yields equalities for the principal curvatures k_j,

$$k_j(k_j - \tau_1) = r \quad \forall j.$$

Hence, each k_j is a root of a quadratic polynomial $P_2(k) = k^2 - \tau_1 k - r$. Its roots are real if and only if $\tau_1^2 + 4r \geq 0$. In the case $r > -\tau_1^2/4$ we have two distinct roots

$$\bar{k}_{1,2} = \frac{1}{2}\left(\tau_1 \pm \sqrt{\tau_1^2 + 4r}\right).$$

Let $n_1 \in (0,n)$ eigenvalues of A are equal to $\bar{k}_1$ and others to $\bar{k}_2$. From $\tau_1 = n_1 \bar{k}_1 + n_2 \bar{k}_2$ and $n = n_1 + n_2$ we obtain

$$n_2 - n_1 = \frac{(n-2)\,\tau_1}{(\tau_1^2 + 4r)^{1/2}} \in \mathbb{Z}. \tag{3.91}$$

If $n_2 = n_1$ then $\tau_1 = 0$ and $k_{1,2} = \pm\sqrt{r}$, otherwise, $\tau_1^2 = \frac{4r}{s^2-1}$ for $s = \frac{n-2}{n_2-n_1}$.
(ii) Assume that μ is constant along the leaves. Then $r = const$ on any leaf. Since $n > 2$, a continuous function $\tau_1 : M \to \mathbb{R}$ has values in a discrete set, hence it is constant. Then all k_j's (from both sets) are constant on M. For n even, (3.91) admits a particular solution: $\tau_1 = 0$, and $n/2$ principal curvatures k_j equal to $\sqrt{r}$, others to $-\sqrt{r}$. $\square$

3.9.3 EGS on Foliated Surfaces

In this section, M^2 is a two-dimensional manifold (a surface) equipped with a transversally orientable foliation $\mathscr{F}$ (by curves) and a 1-parameter family of Riemannian structures (g_t), N a unit normal to $\mathscr{F}$ w.r.t. g_t, and λ_t the geodesic curvature of the leaves w.r.t. N and g_t.

Given $\psi \in C^2(\mathbb{R}^2)$, the EGF g_t of type (b) on $(M^2, \mathscr{F})$ is a solution to the PDE $\partial_t g_t = \psi(\lambda_t, t)\,\hat{g}_t$, where λ_t satisfies the PDE

$$\partial_t \lambda_t + \frac{1}{2} N(\psi(\lambda_t, t)) = 0$$

(see the proof of Proposition 3.4). By Corollary 3.2, if $\psi'_\lambda(\lambda_0, 0) \neq 0$, see the condition (H_1) with $n = 1$, then there is a unique local smooth solution λ_t for $0 \leq t < T$

with initial value λ_0 determined by g_0, and the equality $\hat{g}_t = \hat{g}_0 \exp(\int_0^t \psi'_\lambda(\lambda_t,t)\,dt)$ holds. We have

$$\partial_t(d\,\text{vol}_t) = \frac{1}{2}\psi(\lambda_t,t)\,d\,\text{vol}_t,$$

see (2.21) with $S = \psi(\lambda_t,t)\hat{g}_t$. Hence, for closed M^2, the volume $\text{vol}_t := \int_M d\,\text{vol}_t$ of g_t satisfies the equation

$$\partial_t\,\text{vol}_t = \frac{1}{2}\int_M \psi(\lambda_t,t)\,d\,\text{vol}_t. \tag{3.92}$$

In order to estimate T (i.e., the maximal time interval of Proposition 3.5), suppose that the function ψ does not depend on t and put

$T = \infty$ if $N(\psi(\lambda_0)) \geq 0$ on M, and $T = -2/\inf_M N(\psi(\lambda_0))$ otherwise. Then the EGF $\partial_t g_t = \psi(\lambda_t)\hat{g}_t$ has a unique smooth solution g_t on M^2 for $t \in [0,T)$, and does not possess one for $t \geq T$. If, in addition to condition (H_1), see Sect. 3.4, the inequality $\psi'(\lambda_0)N(\lambda_0) \geq 0$ holds then the solution exists for all $t \geq 0$ (i.e., $T = \infty$).

Proposition 3.11. *The Gaussian curvature K_t of the EGF of type (b) on $(M^2,\mathcal{F},g_t)$ is given by the formula*

$$K_t = \text{div}\left(\exp\left(-\int_0^t \psi(\lambda_t,t)\,dt\right)\nabla^0_N N\right) + N(\lambda_t) - \lambda_t^2. \tag{3.93}$$

Proof. By Lemma 2.10 with $S = \psi(\lambda_t,t)\hat{g}_t$ and $n = 1$, we have

$$\partial_t(\nabla^t_N N) = -\psi(\lambda_t,t)\nabla^t_N N.$$

Integrating the above ODE yields

$$\nabla^t_N N = \exp\left(-\int_0^t \psi(\lambda_t,t)\,dt\right)\nabla^0_N N. \tag{3.94}$$

In our situation here, $K = \text{Ric}(N,N)$, therefore the formula (3.93) is a consequence of (3.94) and Proposition 1.4 with $r = 0$ (see also [54]),

$$\text{div}(\nabla_N N + \tau_1 N) = \text{Ric}(N,N) + \tau_2 - \tau_1{}^2.$$

Indeed, one should apply the identities $\text{div}N = -\tau_1$ and $\text{div}(\tau_1 N) = N(\tau_1) - \tau_1^2$. □

(a) Let $\psi = 1$. The solution to $\partial_t g_t = \hat{g}_t$ is $\hat{g}_t = e^t \hat{g}_0$ $(t \geq 0)$. From $\partial_t \lambda = 0$ we get $\lambda_t = \lambda_0$. By (3.93), the Gaussian curvature is

$$K_t = e^{-t}\,\text{div}(\nabla^0_N N) + N(\lambda_0) - \lambda_0^2.$$

There exists the limit as $t \to \infty$: $K_\infty = N(\lambda_0) - \lambda_0^2$.

(b) Let $\psi = \lambda$, i.e., $\partial_t g = \hat{b}_1$. Then $\lambda_t(s) = \lambda_0(s + \frac{1}{2}t)$ along any N-curve $\gamma(s)$ in the (t,s)-plane. From the EGF's equation

$$\partial_t g_t = \lambda_0(s + t/2)\hat{g}_t$$

we obtain

$$\hat{g}_t(s) = \hat{g}_0(s)\exp\left(\int_0^t \lambda_0(s + \xi/2)\,d\xi\right) \quad (t \in \mathbb{R}).$$

For closed M^2, by (3.92) we have $\mathrm{vol}_t = \mathrm{const}$.

(c) Let $\psi = \lambda^2$. Then

$$\partial_t \lambda + \lambda N(\lambda) = 0 \quad \text{(the Burgers' equation)}.$$

If $N(\lambda_0) \geq 0$ (for $t = 0$) then the solution λ_t exists for all $t \geq 0$.

Example 3.14. Let a function $f \in C^2(-1,1)$ has vertical asymptotes $x = \pm 1$. Consider the foliation $\mathcal{F}$ in the closed strip $\Pi = [-1,1] \times \mathbb{R}$ (equipped with the standard flat metric) whose leaves are

$$L_{\pm} = \{x = \pm 1\}, \quad L_s(x) = \{(x, \, f(x) + s), \, |x| < 1\}, \quad \text{where} \quad s \in \mathbb{R}.$$

The normal N at the origin is directed along y-axis. The tangent and normal to $\mathcal{F}$ unit vector fields (on the whole strip) are

$$X = [\cos\alpha(x), \, \sin\alpha(x)], \quad N = [-\sin\alpha(x), \, \cos\alpha(x)],$$

where $\alpha(x)$ is the angle between the leaves L_s and the x-axis at the intersection points. That is, f and α are related by

$$f'(x) = \tan\alpha(x) \quad \text{and} \quad \cos\alpha = [1 + (f')^2]^{-1/2}, \quad \sin\alpha = f'[1 + (f')^2]^{-1/2}.$$

The curvature of L_s is

$$\lambda_0(x) = f''(x)[1 + (f'(x))^2]^{-3/2} = \alpha'(x) \cdot |\cos\alpha(x)|, \quad |x| < 1.$$

N-curves through the critical points of f are vertical, and divide Π into substrips. Typical foliations in the strip $|x| < 1$ with one vertical trajectory $x = 0$ are the following:

(a) f has exactly one strong minimum at $x = 0$.
(b) f is monotone increasing with one critical point $x = 0$;

Taking

$$f = \frac{1}{10}\left[e^{x^2/(1-x^2)} - 1\right] \quad \text{or} \quad \alpha(x) = \frac{\pi}{2}x$$

for (a), we obtain the Reeb strip, see Fig. 3.4c. For (b) one may take

$$f = \tan\left(\frac{\pi}{2}x\right), \quad \text{or} \quad \alpha(x) = \frac{\pi}{2}x^2.$$

Let $\psi = \psi(\lambda_t)$, where $\lambda_t(x)$ is known for a positive time interval $[0,\varepsilon)$, see Proposition 3.5. We use $X \in T\mathscr{F}$ and normal N to represent the standard frame

$$e_1 = \cos\alpha(x)X - \sin\alpha(x)N, \quad e_2 = \sin\alpha(x)X + \cos\alpha(x)N$$

in the (x,y)-plane. By $\hat{g}_t = \hat{g}_0\, e^{\int_0^t \psi(\lambda_t(x))\,dt}$, we have

$$g_t(X,X) = e^{\int_0^t \psi(\lambda_t(x))\,dt}, \quad g_t(X,N) = 0, \quad g_t(N,N) = 1.$$

The g_t-scalar products of the frame $\{e_1,e_2\}$ are

$$E_t = g_t(e_1,e_1) = \sin^2\alpha + \cos^2\alpha\, e^{\int_0^t \psi(\lambda_t(x))\,dt},$$

$$F_t = g_t(e_1,e_2) = \sin\alpha\cos\alpha[e^{\int_0^t \psi(\lambda_t(x))\,dt} - 1],$$

$$G_t = g_t(e_2,e_2) = \cos^2\alpha + \sin^2\alpha\, e^{\int_0^t \psi(\lambda_t(x))\,dt}.$$

The Gaussian curvature of $g = E\,dx^2 + 2F\,dx\,dy + G\,dy^2$ is given by the known formula

$$K = \frac{-1}{2\sqrt{EG-F^2}}\left(\partial_x\left(\frac{\partial_x G - \partial_y F}{\sqrt{EG-F^2}}\right) + \partial_y\left(\frac{\partial_y E - \partial_x F}{\sqrt{EG-F^2}}\right)\right) - \frac{1}{4(EG-F^2)^2}\begin{vmatrix} E & \partial_x E & \partial_y E \\ F & \partial_x F & \partial_y F \\ G & \partial_x G & \partial_y G \end{vmatrix}$$

which in our case, when g_t do not depend on the y-coordinate, reads:

$$K_t = \frac{-1}{2\sqrt{E_t G_t - F_t^2}}\,\partial_x\left(\frac{\partial_x G_t}{\sqrt{E_t G_t - F_t^2}}\right).$$

We have $E_t G_t - F_t^2 = e^{\int_0^t \psi(\lambda_t(x))\,dt}$ while the N-curves satisfy the ODEs

$$dx/dt = -\sin\alpha(x), \quad dy/dt = \cos\alpha(x).$$

From the first of the ODEs above, we deduce the implicit formula

$$t = -\int_x^{\phi_t(x)} \frac{dx}{\sin\alpha(x)}$$

for local diffeomorphisms $\phi_t(x)$ ($|x| < 1, t \geq 0$) of the N-flow.

Suppose that $\psi(\lambda) = \lambda$. As $\lambda_t(s) = \lambda_0(s + \frac{t}{2})$ is a simple wave along N-curves, we have

$$\lambda_t(x) = \lambda_0(\phi_{t/2}(x)).$$

For example, $\lambda_t(0) = \lambda_0(0)$ for all $t \geq 0$.

Substituting E_t, F_t, and G_t into the above formula for K_t yields

$$K_t = \frac{1}{8}(\cos(2\alpha) - 1)\left(2\int_0^t \lambda_t'' \, dt + \left(\int_0^t \lambda_t' \, dt\right)^2\right) - \left(\cos(2\alpha)(\alpha')^2\right.$$
$$\left. + \frac{1}{2}\sin(2\alpha)\alpha''\right)\left(1 - e^{-\int_0^t \lambda_t \, dt}\right) - \frac{1}{4}\sin(2\alpha)\alpha'\int_0^t \lambda_t' \, dt\left(3 + e^{-\int_0^t \lambda_t \, dt}\right).$$
$$(3.95)$$

Because $\alpha(0) = 0$ and $\lambda_t(0) = \lambda_0(0)$, one has

$$K_t(0) = -(\alpha'(0))^2\left(1 - e^{-t\lambda_0(0)}\right).$$

In case(i), we get $\alpha'(0) > 0$ and $\lambda_0(0) > 0$, so $K_t(0) < 0$ for $t > 0$. As $\lim_{t\to\infty} \phi_t(x) = 0$ for $|x| \leq 1$, there also exists $\lim_{t\to\infty} \lambda_t(x) = \lambda_0(0) > 0$. Hence for any $x \in [-1, 1]$ there is $t_x > 0$ such that $K_t(x) < 0$ for $t > t_x$.

In case (ii) with $\alpha(x) = (\pi/2)x^2$, we have $\alpha'(0) = 0$ and $\alpha''(0) = \pi$. Moreover, $\lambda_0(0) = 0$ and $\lambda_0'(0) = \pi \neq 0$. Since $\lim_{t\to\infty} \phi_t(x) = 0$ for all $0 < x \leq 1$, there also exists

$$\lim_{t\to\infty} \lambda_t(x) = \lim_{t\to\infty} \lambda_0(\phi_{t/2}(x)) = \lambda_0(0) = 0.$$

By (3.95) we have $K_t(0) \equiv 0$ and the series expansion

$$K_t(x) = -\frac{5}{2}\pi^2 x^3 \int_0^t \lambda_t' \, dt + O_t(x^4).$$

We conclude that there exists $t_0 > 0$ such that $K_t(x)$ changes its sign for $t > t_0$ when we cross the line $x = 0$.

Example 3.15. (a) Consider a foliation $\mathscr{F}$ by circles $L_\rho = \{\rho = c\}$ in the annulus $\Omega = \{c_1 \leq \rho \leq c_2\}$ for some $c_2 > c_1 > 0$ with polar coordinates (ρ, θ). Then $X = \partial_\theta$ and $N = \partial_\rho$ are tangent and normal vector fields to the foliation. The standard flat metric is given by:

$$ds^2 = d\rho^2 + G_t(\rho)\,d\theta^2.$$

Notice that $\lambda_0(\rho) = 1/\rho$. Because $\partial_t G_t = \psi(\lambda_t(\rho), t)G_t$, we have

$$G_t = \rho^2 \exp\left(\int_0^t \psi(\lambda_t(\rho), t)\,dt\right).$$

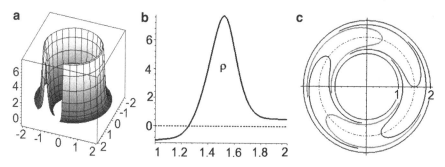

Fig. 3.3 (**a**)–(**b**) Graphs of $\lambda(\rho)$. (**c**) Reeb annulus

From the formula for the Gaussian curvature of Example 3.14 we have

$$K_t = -\frac{1}{2\sqrt{G_t}}\,\partial_\rho\left(\frac{\partial_\rho G_t}{\sqrt{G_t}}\right).$$

Let $\underline{\psi = \lambda}$. Then $\lambda_t(s) = \lambda_0(s+\frac{t}{2})$ on the N-curves. For the foliation by circles $\rho = c$ we have $\lambda_t(\rho) = (\rho+\frac{t}{2})^{-1}$. The Gaussian curvature satisfies $K_t \equiv 0$ for all t.

(b) Consider the *Reeb foliation* in the ring $\Omega = \{1 \le \rho \le 2\}$ of $\mathbb{R}^2$. The leaves L_s ($s \in \mathbb{R}$) and the boundary circles $L_\pm$ are parameterized as

$$L_s(x) = \frac{1}{2}(3+x)[\cos(2\pi(f(x)+s)),\ \sin(2\pi(f(x)+s))], \quad |x| < 1,$$

$$L_\pm(\theta) = \frac{1}{2}(3\pm1)[\cos(2\pi\theta),\ \sin(2\pi\theta)], \quad 0 \le \theta < 1,$$

where, for example, $f = \frac{1}{10}[e^{x^2/(1-x^2)} - 1]$. In polar coordinates we have

$$L_s:\ \rho = \frac{1}{2}(3+x),\ \theta = 2\pi(f(x)+s).$$

The curvature of L_s (at $t = 0$) is

$$k_s(x) = 4\pi\,\frac{4\pi^2(3+x)^2(f'(x))^3 + 2f'(x) + (3+x)f''(x)}{[1+4\pi^2(3+x)^2(f'(x))^2]^{3/2}}, \quad |x| < 1.$$

The unique solution to $k_s(x) = 0$ is $x_0 \approx -0.49$. The function λ of the Reeb foliation is rotationally symmetric, and is given in polar coordinates (θ,ρ) as $\lambda(\theta,\rho) = k_s(2\rho - 3)$. The unique solution to $\lambda(\rho) = 0$ is $\rho_0 = (x_0+3)/2 \approx 1.25$. Hence $\lambda(\rho) < 0$ for $1 \le \rho < \rho_0$ and $\lambda(\rho) > 0$ for $\rho_0 < \rho \le 2$, see Fig. 3.3a–c. Among all N-trajectories, one is closed, $\rho = \rho_0$, others approach to it as $t \to \infty$. Notice that there exists $\lambda_\infty(x) = \lim\limits_{t\to\infty}\lambda_t(x) = k_s(2\rho_0 - 3)$.

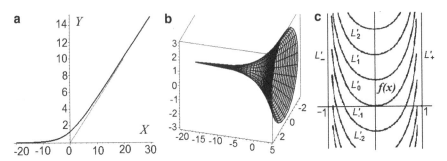

Fig. 3.4 (a) Graph of γ. (b) Hypersurface of revolution (of γ) with $\lambda = const$. (c) Reeb strip

3.9.4 EGS on Hypersurfaces of Revolution

In case of *rotational symmetric metrics* on $M = \mathbb{R} \times S^n$,

$$g = dx_0^2 + \varphi^2(x_0)\, ds_n^2, \quad \text{where } ds_n^2 \text{ is the metric of curvature } 1,$$

the n-parallels $\{x_0 = c\}$ form a Riemannian umbilical foliation $\mathscr{F}$ with the unit normal field $N = \partial_0$. Hence, the EGFs preserve rotational symmetric metrics. (Notice that there are no umbilical foliations with $\lambda = const \neq 0$ on closed, or of finite volume, see [46], manifolds). In this case, the EGFs with generating functions $f_j = f_j(\overrightarrow{\tau})$ can be reduced to (3.76) with ψ of (3.47), and λ_t can be found from (3.48).

Any leaf-wise Killing field $X \perp \partial_0$ provides the EGS structure on M^{n+1} with the rotational symmetric metric g and foliation by parallels. The unit normal field N with g as above is also the EGS structure.

Assuming $\hat{g}_t = \varphi_t^2\, \hat{g}_0$ and using (3.76), we obtain $\lambda_t = -(\varphi_t)_{,0}/\varphi_t^2$ and

$$\partial_t \varphi_t = \frac{1}{2}\,\psi(\lambda_t)\,\varphi_t \quad \Rightarrow \quad \varphi_t = \varphi_0 \exp\left(\frac{1}{2}\int_0^t \psi(\lambda_t)\, dt\right).$$

In the particular case of $\psi(\lambda) = \lambda$, we get the linear PDE $\partial_t \lambda + \frac{1}{2}N(\lambda) = 0$ representing the "unidirectional wave motion" along any N-curve $\gamma(s)$,

$$\lambda_t(s) = \lambda_0(s - t/2). \tag{3.96}$$

If $\lambda_0 = c \in \mathbb{R}$ then $\lambda_t = c$ and $\varphi_t = \varphi_0 \exp(\frac{t}{2}\,\psi(c))$. Rotationally symmetric metrics with $\lambda = const$ exist on hyperbolic space $\mathbb{H}^{n+1}$ with horosphere foliation, Fig. 3.2.

For example, assume $n = 1$, and consider $(M^2, g_t, \mathscr{F})$ in biregular foliated coordinates (x_0, x_1). Hence $(g_t)_{00} = 1$ and $(g_t)_{11} = \varphi_t^2$. By Lemma 2.2 for $b_{11} = g_t(A\partial_1, \partial_1) = \lambda_t\, \varphi_t^2$, we have $\lambda_t = -(\varphi_t)_{,0}/\varphi_t^2$. The Gaussian curvature of M^2 is

$$K_t = -(\varphi_t)_{,00}/\varphi_t.$$

Example 3.16 (EGS on hypersurfaces of revolution). Some of rotationally symmetric metrics come from hypersurfaces of revolution in space forms. Evolving
them by EGF corresponding to the foliation by parallels yields deformations of
hypersurfaces of revolution. Revolving the graph of $x_1 = f(x_0)$ about the x_0-axis
of $\mathbb{R}^{n+1}$, we get the hypersurface $M^n : f^2(x_0) = \sum_{i=1}^n x_i^2$ foliated by $(n-1)$-spheres
$\{x_0 = c\}$ (parallels) with the induced metric

$$g = (1 + f'(x_0)^2)\,dx_0^2 + f^2(x_0)\sum_{i=1}^n dx_i^2. \qquad (3.97)$$

(a) Revolving a line $\gamma_0 : x_1 = x_0 \tan \beta$ about the x_0-axis, we build the cone $C_0 :$
$(x_0 \tan \beta)^2 = \sum_{i=1}^n x_i^2$, with the metric

$$g_0 = dx_0^2 + (x_0 \sin \beta)^2 \sum_{i=1}^n dx_i^2.$$

Hence

$$\varphi_0 = x_0 \sin \beta, \qquad \lambda_0(x_0) = -2/x_0.$$

Applying the EGF $\partial_t g_t = \lambda_t \hat{g}_t$, we obtain by (3.96) that

$$\lambda_t(x_0) = -\frac{2}{x_0 - t/2}.$$

The rotationally symmetric metric

$$g_t = dx_0^2 + \left(x_0 - \frac{t}{2}\right)^2 (\sin^2 \beta)\sum_{i=1}^n dx_i^2$$

appears on the same cone translated across the x_0-axis,

$$C_t : \left(x_0 - \frac{t}{2}\right)^2 \tan^2 \beta = \sum_{i=1}^n x_i^2.$$

Any leaf-wise Killing field $X \perp N$ with the induced metric g provide an EGS
structure on M^n.
(b) Let us find a curve $y = f(x) > 0$ such that the metric (3.97) on the hypersurface
of revolution $M^n : \sum_{i=1}^n x_i^2 = f^2(x_0)$ has $\lambda_0 = \text{const} = 1$. Using $\lambda_0 = \frac{1}{f(x_0)}\sin \phi$,
where $\tan \phi = f'(x_0)$, we obtain the ODE

$$\frac{|f'|}{f\sqrt{1 + (f')^2}} = 1 \Rightarrow \frac{df}{dx} = \frac{f}{\sqrt{4 + f^2}},$$

whose solution is

$$\gamma : x = \log \frac{\sqrt{4 + f^2} - 2}{\sqrt{4 + f^2} + 2} + \sqrt{4 + f^2} + C, \quad \text{where} \quad C \in \mathbb{R}.$$

The hypersurface $M^n \subset \mathbb{R}^{n+1}$ looks like a pseudosphere (for $n = 2$ see Fig. 3.4a,b), however for $x_0 \to \infty$ it is asymptotic to the cone $(x_0 + C)^2 = \sum_{i=1}^{n} x_i^2$. The sectional curvature K satisfies

$$K(\partial_0, \partial_1) = -(x_1^2 + 2)^{-2} < 0, \qquad \lim_{x_1 \to \pm\infty} K = 0.$$

As for a horosphere foliation on hyperbolic space, the unit normal N to parallels and any leaf-wise Killing field $X \perp N$ provide EGS structures on (M^n, g).

References

1. L.J. Alias, S. de Lira, J.M. Malacarne: Constant higher-order mean curvature hypersurfaces in Riemannian spaces, J. Inst. of Math. Jussieu, 5(4), (2006) 527–562.
2. B. Andrews B., C. Hopper: The Ricci Flow in Riemannian Geometry, LNM 2011, Springer, 2011, 296 pp.
3. K. Andrzejewski and P. Walczak: The Newton transformation and new integral formulae for foliated manifolds, Ann. Glob. Anal. Geom. 37 (2) (2009), 103–111.
4. D. Asimov: Average gaussian curvature of leaves of foliations, Bull. Amer. Math. Soc. 84(1) (1978), 131–133.
5. P. Baird P., J. Wood: Harmonic morphisms between Riemannian manifolds. London Math. Soc. Monographs 29, Oxford University Press, 2003, 520 pp.
6. J.L.M. Barbosa, K. Kenmotsu, G. Oshikiri: Foliations by hypersurfaces with constant mean curvature. Math. Z. 207, (1991) 97–108.
7. L. Bessierès et al: Geometrization of 3-manifolds, EMS, 2010, 237 pp.
8. S. Brendle: Ricci Flow and the Sphere Theorem, Graduate Studies in Math., 111, AMS, 2010, 176 pp.
9. F. Brito, R. Langevin, H. Rosenberg: Intégrales de courbure sur des variétés feuilletées, J. Diff. Geom. 16 (1981), 19–50.
10. F. Brito and P. Walczak: On the energy of unit vector fields with isolated singularities. Annales Polonici Math. LXXIII 316 (2000), 269–274.
11. M. Brunella and E. Ghys: Umbilical foliations and transversely holomorphic flows, J. Diff. Geom. 41 (1995), No 3, 1–19.
12. A. Candel and L. Conlon: Foliations, I, II, AMS, Providence, 2000, 2003.
13. B. Chow and D. Knopf: The Ricci Flow: An Introduction, AMS, 2004.
14. B. Chow et al: The Ricci Flow: Techniques and Applications, Parts I, II, III, AMS, 2007, 2010.
15. T. Cecil and S. Chern (eds.): Tight and taut submanifolds. Papers in memory of Nicolaas H. Kuiper, Cambridge University Press, 1997.
16. M. Czarnecki and P. Walczak: Extrinsic geometry of foliations, 149–167, in "Foliations 2005", World Scientific Publication, NJ, 2006
17. W. Chen and J. Louck: The Combinatorial Power of the Companion Matrix, Linear Algebra and Its Applications 232 (1996) 261–278.
18. K. Ecker: Regularity theory for Mean curvature flow. Birkhäuser, Boston, 2004, 165 pp.
19. J. Eells and J. Sampson: Harmonic Mappings of Riemannian Manifolds, Amer. J. Math. 86, 109–160 (1964).
20. A. Epstein and E. Vogt: A counterexample to the periodic orbit conjecture in codimension 3. Ann. of Math. (2-nd series) 108(3) (1978), 539–552.

V. Rovenski and P. Walczak, *Topics in Extrinsic Geometry of Codimension-One Foliations*, SpringerBriefs in Mathematics, DOI 10.1007/978-1-4419-9908-5,
© Vladimir Rovenski and Paweł Walczak 2011

21. K. Gerhardt: Curvature Problems. Series in Geometry and Topology, 39, Int. Press, 2006, 323 pp.
22. C. Godbillon: Dynamical Systems on Surfaces, Springer Verlag, 1983.
23. R. Hamilton: Three-Manifolds with Positive Ricci Curvature, J. Diff. Geom. 17, 255–306 (1982).
24. P. Hartman and A. Wintner: On Hyperbolic Partial Differential Equations, Amer. J. Math. 74 (1952) 834–864.
25. B. Kleiner and J. Lott: Notes on Perelmans papers, Geom. and Topology 12 (2008) 2587–2855.
26. R. Langevin: Feuilletages tendus. Bull. Soc. Math. France 107 (1979), 271–281.
27. R. Langevin and G. Levitt: Sur la courbure totale des feuilletages des surfaces a bord, Bol. Soc. Brasil. Mat. 16 (1985), 1–13.
28. R. Langevin and C. Possani: Total curvature of foliations, Illinois J. Math. 37 (1993), 508–524.
29. R. Langevin and H. Rosenberg: Fenchel type theorems for submanifolds of S^n, Comment. Math. Helv. 71 (1996), 594–616.
30. R. Langevin and P. Walczak: Conformal geometry of foliations. Geom. Dedicata 132 (2008), 135–178.
31. L. Nirenberg: Topics in Nonlinear Functional Analysis. AMS, 2001.
32. S. Novikov: Topology of foliations, Trudy Moskov. Mat. Obsc. 14 (1965), 248–278.
33. G. Oshikiri: A characterization of the mean curvature functions of codimension-one foliations. Tohoku Math. J. 49 (1997), 557–563.
34. ———: Some properties of mean curvature vectors for codimension-one foliations. Illinois J. of Math. 49 (1), (2005), 159–166.
35. G. Perelman: The entropy formula for the Ricci flow and its geometric applications, arXiv.org/math.DG/0211159 v1 (2002).
36. ———: Ricci flow with surgery on three-manifolds, arXiv.org/math.DG/0303109,Ĭ (2003).
37. ———: Finite extinction time for the solutions to the Ricci flow on certain 3-manifolds, arXiv.org/math.DG/0307245 v1 (2003).
38. R. Ponge, and H. Reckziegel: Twisted products in pseudo-Riemannian geometry, Geom. Dedicata 48 (1993), 15–25.
39. A. Ranjan: Structural equations and an integral formula for foliated manifolds, Geom. Dedicata 20 (1986), 85–91
40. G. Reeb: Sur certaines propriétés topologiques des variétés feuilletées, Actualités, Sci. Ind., no. 1183, Hermann & Cie., Paris, 1952.
41. ———: Sur la courboure moyenne des variétés intégrales d'une équation de Pfaff $\omega = 0$, C. R. Acad. Sci. Paris 231 (1950), 101–102.
42. V. Rovenski: Foliations on Riemannian Manifolds and Submanifolds, Birkhäuser, 1998.
43. ———: Foliations, submanifolds and mixed curvature, J. Math. Sci. New York, (2000), 99(6), 16991787.
44. ———: Integral formulae for a Riemannian manifold with two orthogonal distributions, CEJM, 2011, DOI 10.2478/s11533-011-0026-y.
45. V. Rovenski, and P. Walczak: Integral formulae on foliated symmetric spaces, Mathematische Annalen, 2011, DOI 10.1007/s00208-011-0637-4.
46. ———: Integral fomulae for foliations on Riemannian manifolds, in Proc. of 10-th Int. Conf. "Diff. Geometry and Its Applications", Olomouc, 193–204, World Scientific, 2008.
47. ———: Variational formulae for the total mean curvatures of a codimension-one distribution. Proc. of 8-th Int. Colloq., Santiago-de Compostela, Spain, July 2008, 83–93, World Scientific, 2009.
48. P. Schweitzer and P. Walczak: Prescribing mean curvature vectors for foliations. Illinois J. Math. 48 (2004), 21–35.
49. D. Sullivan: Cycles for the dynamical study of foliated manifolds and complex manifolds, Invent. Math. 36 (1976), 225–256.
50. ———: A homological characterization of foliations consisting of minimal surfaces. Comm. Math. Helv. 54 (1979), 218–223.

51. M. Taylor: Partial Differential Equations, III: Nonlinear Equations, 2nd Ed., Applied Math. Sciences 117, Springer 2011.
52. P. Tondeur: Geometry of foliations, Birkhäuser, Verlag, Basel, 1997
53. P. Topping: Lectures on the Ricci flow, LMS Lecture Notes 325, London Math. Society and Cambridge University Press, 2006.
54. P. Walczak: An integral formula for a Riemannian manifold with two orthogonal complementary distributions. Colloq. Math. 58 (1990), 243–252.
55. ———— : Mean curvature invariant foliations, Illinois J. Math. 37 (1993), 609–623.
56. ———— : Dynamics of foliations, groups and pseudogroups, Birkhäuser, Basel, 2004.
57. ———— : On foliations with leaves satisfying some geometrical conditions, Dissertationes Math. 226 (1983), 1–51.
58. ———— : Mean curvature functions for codimension-one foliations with all the leaves compact, Czech. Math. J. 34 (1984), 146–155.
59. S. Walczak: Warped compact foliations, Annales Polonici Math. 94(3) (2008), 231–243.

Index

V. Rovenski and P. Walczak, *Topics in Extrinsic Geometry of Codimension-One Foliations*, SpringerBriefs in Mathematics, DOI 10.1007/978-1-4419-9908-5,
© Vladimir Rovenski and Paweł Walczak 2011